CAMBRIDGE LIBRARY COLLECTION

Books of enduring scholarly value

Botany and Horticulture

Until the nineteenth century, the investigation of natural phenomena, plants and animals was considered either the preserve of elite scholars or a pastime for the leisured upper classes. As increasing academic rigour and systematisation was brought to the study of 'natural history', its subdisciplines were adopted into university curricula, and learned societies (such as the Royal Horticultural Society, founded in 1804) were established to support research in these areas. A related development was strong enthusiasm for exotic garden plants, which resulted in plant collecting expeditions to every corner of the globe, sometimes with tragic consequences. This series includes accounts of some of those expeditions, detailed reference works on the flora of different regions, and practical advice for amateur and professional gardeners.

The Trees of Great Britain and Ireland

Although without formal scientific training, Henry John Elwes (1846–1922) devoted his life to natural history. He had studied birds, butterflies and moths, but later turned his attention to collecting and growing plants. Embarking on his most ambitious project in 1903, he recruited the Irish dendrologist Augustine Henry (1857–1930) to collaborate with him on this well-illustrated work. Privately printed in seven volumes between 1906 and 1913, it covers the varieties, distribution, history and cultivation of tree species in the British Isles. The strictly botanical parts were written by Henry, while Elwes drew on his extensive knowledge of native and non-native species to give details of where remarkable examples could be found. Each volume contains photographic plates as well as drawings of leaves and buds to aid identification. The species covered in Volume 1 (1906) include beech, spruce and yew.

Cambridge University Press has long been a pioneer in the reissuing of out-of-print titles from its own backlist, producing digital reprints of books that are still sought after by scholars and students but could not be reprinted economically using traditional technology. The Cambridge Library Collection extends this activity to a wider range of books which are still of importance to researchers and professionals, either for the source material they contain, or as landmarks in the history of their academic discipline.

Drawing from the world-renowned collections in the Cambridge University Library and other partner libraries, and guided by the advice of experts in each subject area, Cambridge University Press is using state-of-the-art scanning machines in its own Printing House to capture the content of each book selected for inclusion. The files are processed to give a consistently clear, crisp image, and the books finished to the high quality standard for which the Press is recognised around the world. The latest print-on-demand technology ensures that the books will remain available indefinitely, and that orders for single or multiple copies can quickly be supplied.

The Cambridge Library Collection brings back to life books of enduring scholarly value (including out-of-copyright works originally issued by other publishers) across a wide range of disciplines in the humanities and social sciences and in science and technology.

The Trees
of Great Britain
and Ireland

VOLUME 1

HENRY JOHN ELWES
AUGUSTINE HENRY

CAMBRIDGE
UNIVERSITY PRESS

CAMBRIDGE
UNIVERSITY PRESS

University Printing House, Cambridge, CB2 8BS, United Kingdom

Published in the United States of America by Cambridge University Press, New York

Cambridge University Press is part of the University of Cambridge.
It furthers the University's mission by disseminating knowledge in the pursuit of
education, learning and research at the highest international levels of excellence.

www.cambridge.org
Information on this title: www.cambridge.org/9781108069328

© in this compilation Cambridge University Press 2014

This edition first published 1906
This digitally printed version 2014

ISBN 978-1-108-06932-8 Paperback

Selected botanical reference works available in the
CAMBRIDGE LIBRARY COLLECTION

al-Shirazi, Noureddeen Mohammed Abdullah (compiler), translated by Francis Gladwin: *Ulfáz Udwiyeh, or the Materia Medica* (1793) [ISBN 9781108056090]

Arber, Agnes: *Herbals: Their Origin and Evolution* (1938) [ISBN 9781108016711]

Arber, Agnes: *Monocotyledons* (1925) [ISBN 9781108013208]

Arber, Agnes: *The Gramineae* (1934) [ISBN 9781108017312]

Arber, Agnes: *Water Plants* (1920) [ISBN 9781108017329]

Bower, F.O.: *The Ferns (Filicales)* (3 vols., 1923–8) [ISBN 9781108013192]

Candolle, Augustin Pyramus de, and Sprengel, Kurt: *Elements of the Philosophy of Plants* (1821) [ISBN 9781108037464]

Cheeseman, Thomas Frederick: *Manual of the New Zealand Flora* (2 vols., 1906) [ISBN 9781108037525]

Cockayne, Leonard: *The Vegetation of New Zealand* (1928) [ISBN 9781108032384]

Cunningham, Robert O.: *Notes on the Natural History of the Strait of Magellan and West Coast of Patagonia* (1871) [ISBN 9781108041850]

Gwynne-Vaughan, Helen: *Fungi* (1922) [ISBN 9781108013215]

Henslow, John Stevens: *A Catalogue of British Plants Arranged According to the Natural System* (1829) [ISBN 9781108061728]

Henslow, John Stevens: *A Dictionary of Botanical Terms* (1856) [ISBN 9781108001311]

Henslow, John Stevens: *Flora of Suffolk* (1860) [ISBN 9781108055673]

Henslow, John Stevens: *The Principles of Descriptive and Physiological Botany* (1835) [ISBN 9781108001861]

Hogg, Robert: *The British Pomology* (1851) [ISBN 9781108039444]

Hooker, Joseph Dalton, and Thomson, Thomas: *Flora Indica* (1855) [ISBN 9781108037495]

Hooker, Joseph Dalton: *Handbook of the New Zealand Flora* (2 vols., 1864–7) [ISBN 9781108030410]

Hooker, William Jackson: *Icones Plantarum* (10 vols., 1837–54) [ISBN 9781108039314]

Hooker, William Jackson: *Kew Gardens* (1858) [ISBN 9781108065450]

Jussieu, Adrien de, edited by J.H. Wilson: *The Elements of Botany* (1849) [ISBN 9781108037310]

Lindley, John: *Flora Medica* (1838) [ISBN 9781108038454]

Müller, Ferdinand von, edited by William Woolls: *Plants of New South Wales* (1885) [ISBN 9781108021050]

Oliver, Daniel: *First Book of Indian Botany* (1869) [ISBN 9781108055628]

Pearson, H.H.W., edited by A.C. Seward: *Gnetales* (1929) [ISBN 9781108013987]

Perring, Franklyn Hugh et al.: *A Flora of Cambridgeshire* (1964) [ISBN 9781108002400]

Sachs, Julius, edited and translated by Alfred Bennett, assisted by W.T. Thiselton Dyer: *A Text-Book of Botany* (1875) [ISBN 9781108038324]

Seward, A.C.: *Fossil Plants* (4 vols., 1898–1919) [ISBN 9781108015998]

Tansley, A.G.: *Types of British Vegetation* (1911) [ISBN 9781108045063]

Traill, Catherine Parr Strickland, illustrated by Agnes FitzGibbon Chamberlin: *Studies of Plant Life in Canada* (1885) [ISBN 9781108033756]

Tristram, Henry Baker: *The Fauna and Flora of Palestine* (1884) [ISBN 9781108042048]

Vogel, Theodore, edited by William Jackson Hooker: *Niger Flora* (1849) [ISBN 9781108030380]

West, G.S.: *Algae* (1916) [ISBN 9781108013222]

Woods, Joseph: *The Tourist's Flora* (1850) [ISBN 9781108062466]

For a complete list of titles in the Cambridge Library Collection please visit:
www.cambridge.org/features/CambridgeLibraryCollection/books.htm

THE TREES OF GREAT BRITAIN AND IRELAND

QUEEN BEECH AT ASHRIDGE

From a Drawing lent by the Earl Brownlow.

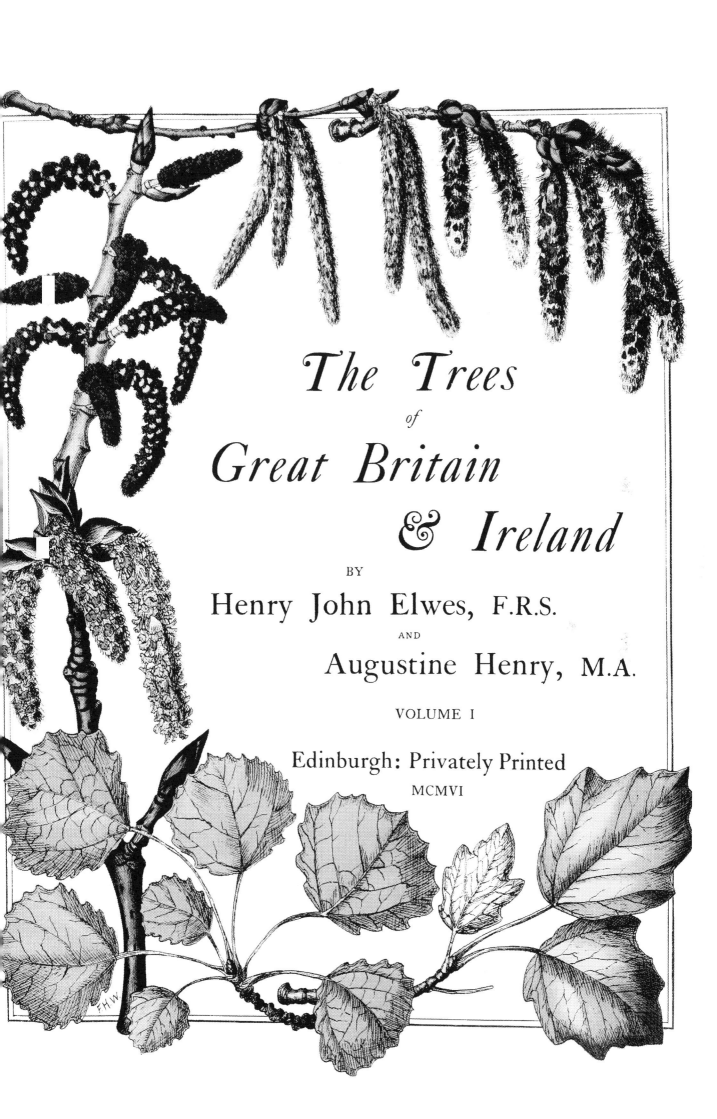

The Trees

of

Great Britain

& Ireland

BY

Henry John Elwes, F.R.S.

AND

Augustine Henry, M.A.

VOLUME I

Edinburgh: Privately Printed

MCMVI

DEDICATED

BY SPECIAL PERMISSION

TO

𝕳𝖎𝖘 𝕸𝖆𝖏𝖊𝖘𝖙𝖞 𝕶𝖎𝖓𝖌 𝕰𝖉𝖜𝖆𝖗𝖉 VII

BY

HIS OBEDIENT HUMBLE SERVANTS

THE AUTHORS

PREFACE

THE United Kingdom offers a hospitality to exotic vegetation which finds no parallel in the Northern Temperate region of the globe. Never parched by the heat of a continental summer, the rigour of winter is no less tempered by its insular position. The possession of land still ensures the residence on their properties of a large number of persons of at least moderate affluence. The most modest country house possesses a garden, and not rarely some sort of pleasure ground; and this usually reaches the dimensions of a park in the case of the larger mansions. While forests for the commercial production of timber such as are found in foreign countries hardly exist, and the methods of their scientific management are little recognised, arboriculture of some sort may be almost said to be a national passion. In all but purely agricultural districts the free and unrestrained growth of trees enhances, if it does not create, the natural beauty of the landscape. The Roman occupation brought to our shores our fruit-trees and others whose names of Latin derivation bear witness to their foreign origin. One of these, the so-called "English Elm," dominates the landscape of Southern England. Yet, while it perfects its seed on the Continent, it rarely does so in this country, and it holds its own by root suckers, the tenacity of which is all but ineradicable.

Down to the reign of Henry the Eighth the native forests supplied the timber necessary for construction. It was not till their area became restricted that planting was commenced to maintain the supply. And if this has never developed into a scientific system as it has done abroad, the reason may be found in the abandonment of wood as fuel for coal, and the facilities for external supply of over-sea water-carriage which attach to a maritime country.

From an early time with the growth of continental intercourse, the contents of foreign gardens had gradually been transferred to those of the wealthy at home. The taste, however, for cultivating foreign trees and shrubs simply for their interest, and apart from any useful purpose they might serve, is not more recent than the seventeenth century. The pioneer in this branch of English arboriculture was Henry Compton, Bishop of London, who planted in the garden of Fulham Palace "a greater variety of curious exotic plants and trees than had at that time been collected in any garden in England." Hitherto the European continent had been the only hunting ground. To this was now added in striking contrast the resources of the North American forests.

In the eighteenth century the practice of planting foreign trees became in some degree a fashion amongst wealthy landowners, though still mainly for ornament. This was due in large measure to the example of Archibald, third Duke of Argyll, who formed a large collection at Whitton. After his death in 1762 all that were removable were transferred to Kew, where an Arboretum had been commenced by the Princess Dowager of Wales.

An intelligent taste for arboriculture was at any rate for a time firmly established. Those who care to trace its further history more in detail will find abundant information in Loudon's *Arboretum et Fruticetum Britannicum*, a work which, though published more than half a century ago, must always remain indispensable to any student of the subject. Parks and pleasure grounds throughout the country were stocked with specimens of new and interesting trees. And though often neglected and even forgotten, we now possess a wealth of examples which have attained adult development. Loudon catalogued with indefatigable industry every tree or shrub known to be tolerant of the climate of the British Isles. It might have been thought that this laborious undertaking would have excited a new interest in planting. But it began to languish with the beginning of the last century, and Loudon's labours from their very completeness, perhaps, deterred many from engaging in an occupation where more than moderate success would seem costly and laborious, and anything beyond almost unattainable. In 1845 a National Arboretum was projected at Kew, and commenced the following year on a plan prepared by W. A. Nesfield.

The latter half of the last century saw a remarkable development of open-air horticulture. In so far as this included woody plants, it was limited to shrubs. Broad-leaved trees were little cared for. The rarer kinds were little in request, and those that were planted were too often drawn from the ill-named stock of some convenient nursery. The neglect was increased when conifers became a fashion. This led, no doubt, to many fine Pinetums being planted, the interest and importance of which will increase with age. But it led also to much unconsidered and scattered planting of trees which, attractive enough in a juvenile state, are often less sightly as they grow older, and can never blend with their broad-leaved neighbours into stately umbrageous masses.

If the planting of broad-leaved trees as distinguished from conifers has for the moment fallen into neglect, we still inherit the results of the labours of our predecessors. The British Isles for the last two centuries have, in fact, been the seat of an experiment in arboriculture without parallel elsewhere. And the very neglect into which tree-planting has fallen, paradoxical as it may seem, adds to the interest and value of the experiment. For the trees that have come down to us from the past have been subjected to the least favourable conditions, and have had in effect to survive a somewhat rigorous process of natural selection.

In taking stock of the results, the task which my friend Mr. Elwes has set himself differs, if I understand his intention rightly, somewhat widely from that which Loudon accomplished. That amounted to little more than a descriptive catalogue of every woody plant the cultivation of which had been attempted in this country. The present work aims at ascertaining the practical results. What are the most favourable conditions for the growth of each species? What in turn are the most suited for different circumstances? And what, if any, profit can be derived from their cultivation on a large scale?

And to accomplish or even attempt such a task appears to me no small public service. The depreciation of the value of agricultural land has turned the thoughts of many landowners to the possibility of growing timber as a crop for profit. So far little attempt has been made to depart from tradition. Yet it cannot be doubted that there must be many trees suited to our climate whose commercial possibilities are still unascertained. Apart from the larger uses of timber its employment for minor industries is still little regarded among us. If but a single tree can be added to the list of those which can be profitably cultivated the labour spent on the quest will not be unrequited.

W. T. THISELTON-DYER.

KEW, *November* 1905.

CONTENTS

ILLUSTRATIONS

INTRODUCTION

THE object of this work is to give a complete account of all the trees which grow naturally or are cultivated in Great Britain, and which have attained, or seem likely to attain, a size which justifies their being looked on as timber trees; but does not include those which are naturally of shrubby or bushy habit.

Although sixty years have passed since Loudon's great work was published, no book has been written which describes in full the species which in his time were unknown, or so recently introduced that their cultural requirements and economic or ornamental value had not been tested.

Many deciduous trees, which were commonly planted before his time, have gone out of popular favour, and are almost forgotten; whilst the rage for conifers which sprang up about seventy years ago has led to the introduction of almost every species which can be grown in this country; and many of these have now reached an age at which their value can be accurately judged of.

Special books dealing with conifers have appeared which may satisfy the wants of a horticulturist, but none exists that at all meets the requirements of landowners, foresters, and arboriculturists, and will enable them to distinguish the species with certainty, or guide them in selecting the species the best suited for economic culture in different parts of England.

Forestry is at last making headway as a science in this country, but too many of the books recently published on the subject have been based on continental experience, which is not directly applicable to the very different conditions of climate, soil, labour, and market existing here. In the cultural part of the work we base our conclusions on home experience and practice; and in this connection it may be stated that for almost every exotic species there are older specimens of individual trees and of plantations in these islands than on the Continent. After having seen the trees of every country in Europe, of nearly all the States of North America, of Canada, Japan, China, West Siberia, and Chile, we confidently assert that these islands contain a greater number of fine trees from the temperate regions of the world than any other country. Descriptions of the best examples of all of these and of interesting woods and plantations will be a prominent feature of the book.

We have the special qualification that we have seen with our own eyes and studied on the spot, both at home and abroad, most of the trees which will be included in the book.

Knowing how difficult it is for the general public to understand the descriptions of nearly allied species, usually made by compilers who are unacquainted with the crucial points of distinction, we hope to supply this information in concise, clear, and simple language. What we understand by scientific knowledge is accuracy, expressed in plain words; and in order to ensure this we have copied nothing from other authors that we could verify for ourselves.

In order to give a history of the finest trees in this country, we have visited during the past five years nearly every important place in England, Scotland, Wales, and Ireland where large and rare trees are found; and have received from land-owners, estate agents, foresters, and gardeners an amount of information and assistance which justifies us in believing that our work will be generally appreciated. Though the historic trees of some places in England and of more in Scotland have been described in scattered publications, those of Ireland have been almost totally neglected; and Dr. Henry has paid special attention to the many interesting properties in that country.

A prominent feature of the work will be the illustrations. Modern photography enables the authors to give accurate pictures of the trees as they grow. Almost all the photographs of trees and of forest scenes have been taken by skilful photographers specially engaged for the purpose. In dealing with about 300 species of trees, many of which will require several illustrations to show the best specimens both as park and forest trees, the authors have accumulated a large number of photographs, which are being reproduced by the Autotype Company of London, who guarantee their permanency.

With regard to these illustrations we desire to say, that though in some cases they may not be perfect from the point of view of the photographic artist, yet the amount of time, skill, and money that has been spent on them is very far beyond what would be imagined by any one who has not had experience of the difficulty of securing good negatives of trees scattered over so large an area, under all conditions of light and weather, and in situations often extremely difficult to the photographer.

In some cases two or three special journeys have been made to obtain a photograph of one tree only, as the object has been to show the finest individual trees known to the authors rather than to make pretty pictures of scenery.

Besides these reproductions of photographs there will be lithographed drawings of seedlings, buds, leaves, flowers, and fruit, so far as is necessary to distinguish the trees in winter and in summer. These original drawings have all been done under the personal supervision of Dr. Henry, who has carefully studied the material, living and dead, that exists in the unrivalled establishment at Kew.

All measurements have been taken by the authors themselves with Stanley's Apomecometer, or by practical foresters on whose accuracy they could rely, and though in many cases errors to the extent of a few feet may have been made, owing to the shape or position of the tree measured, we believe them to be as accurate as possible under the conditions.

FAGUS

THE NORTHERN BEECHES

Fagus, Linnæus, *Syst.* ed. 1. *V. Monœcia* (1735); Bentham et Hooker, *Gen. Plant.* iii. 410 (1880).

THE genus, as understood by Bentham and Hooker, included all the beeches, those of the southern as well as of the northern hemisphere. Blume[1] separated the southern beeches as a distinct genus, Nothofagus; and his arrangement, on account of its convenience, will be followed by us. Fagus belongs to the family Quercineæ, which includes the oaks, chestnuts, castanopsis, and beeches. The genus, limited to include only the northern beeches, consists of large trees with smooth bark and spindle-shaped buds arranged alternately on the twigs in two rows. Leaves: deciduous, simple, pinnately-nerved, folded in the bud along the primary nerves. Flowers monœcious: the staminate flowers numerous in pendulous globose heads, the pistillate flowers in pairs in involucres. The male flower has a 4 to 8 lobed calyx with 8 to 16 stamens. The female flower has a 6 lobed calyx, adnate to a 3 celled ovary, with 2 ovules in each cell; styles 3, filiform. On ripening, the involucre is enlarged, woody, and covered with bristly deltoid or foliaceous processes; it dehisces by 4 valves, allowing the 2 fruits enclosed to escape. Each fruit is 3 angled and contains 1 seed, which has no albumen.

Seven distinct species of Fagus have been described, of which three, the European beech, the American beech, and the peculiar *Fagus japonica* are recognised by all botanists as good species. The Caucasian beech, the two Chinese beeches, and the common beech of Japan are considered by some authorities to be mere varieties of *Fagus sylvatica*; but these can all readily be distinguished, and in the following account will be treated as independent species.

KEY TO THE SPECIES OF FAGUS.

I. *Nuts projecting out of the top of the involucre.*

1. **Fagus japonica.** Japan.
   Involucre very small, covered externally with small deltoid processes, and borne on a very long slender stalk. Leaves with 10-14 pairs of nerves, which bend round before quite reaching the slightly undulating margin.

---

[1] Blume, in *Mus. Lugd. Bat.* i. 306.

II. *Nuts enclosed in the involucre.*

   A. *Involucres with linear, awl-shaped, bristly appendages.   Species 2, 3, and 4.*

     2. **Fagus sylvatica.**   Europe.
     Fruit-stalks short and pubescent throughout.
     Leaves : under surface glabrous except on the nerves and midrib ; lateral
        nerves 5-9 pairs ; margin not regularly serrate.

     3. **Fagus ferruginea.**   North America.
     Fruit-stalks short and pubescent throughout.
     Leaves : under surface glabrous except on the nerves and midrib ; lateral
        nerves 10-12 pairs, ending in the teeth ; margin serrate.

     4. **Fagus sinensis.**   Central China.
     Fruit-stalks short, pubescent only close to the involucre.
     Leaves : minutely pubescent over their whole under surface ; lateral nerves
        9-10 pairs ending in the teeth ; margin serrate.

   B. *Involucres with their lower appendages dilated and foliaceous.   Species 5, 6, 7.*

     5. **Fagus orientalis.**   Caucasus, Asia Minor, N. Persia.
     Fruit-stalks long (twice the length of the involucre or more) and very
        pubescent throughout.
     Leaves : broadest above the middle ; lateral nerves about 10 pairs, bending
        round before quite reaching the undulate margin ; under surface glabrous
        except on the midrib and nerves.

     6. **Fagus Sieboldi.**   Japan.
     Fruit-stalks short (as long as the involucres) and pubescent throughout.
     Leaves : broadest below the middle ; lateral nerves 7-10 pairs, bending round
        before quite reaching the margin, which is crenate ; under surface glabrous
        beneath except on the nerves and midrib.

     7. **Fagus Engleriana.**   Central China.
     Fruit-stalks very long (five times the length of the involucre) and quite glabrous.
     Leaves glabrous and glaucescent underneath ; lateral nerves 13 pairs, bend-
        ing round before quite reaching the undulate margin.

F AGUS FERRUGINEA.   American Beech.

Fagus ferruginea, Dryander, in *Ait. Hort. Kew.* iii. 362 (1789) ; Loudon, *Arb. et Frut.* iii. 1980
   (1838) ; Mayr. *Wald. von Nordamerika*, 176 (1890).
Fagus sylvatica atropunicea, Marsh. *Arb. Am.* 46 (1785).
Fagus silvestris, Mich. fil. *Hist. Arb. Am.* ii. 170, t. 8 (1812).
Fagus atropunicea, Sudworth, *Bull. Torrey Bot. Club*, xx. 42 (1893).
Fagus americana, Sweet, *Hort. Brit.* 370 (1826) ; Sargent, *Silva of N. Am.* ix. 27 (1896).

The American beech ranges, according to Sudworth, from Nova Scotia to
north shore of Lake Huron and Northern Wisconsin ; south, to western Florida ;
and west, to south-eastern Missouri and Texas (Trinity River).   Mayr[1] says it is at

---

[1] Mayr, *l.c.*

its best in the northern deciduous forest, where it is a stately tree, *e.g.* at Lake Superior. The finest individual trees occur on the small hills of the Mississippi valley, but the timber is not so good as that of trees farther north. Pure woods of American beech rarely if ever occur.[1] Elwes saw the American beech principally near Boston and in Canada, and remarked one peculiarity which may not be found in all places. This was its tendency to throw up suckers from the roots, a feature which is very marked in Professor Sargent's park at Brookline, and in the beautiful grounds of the Arnold Arboretum. There is a group of beech here by the side of a drive, of which the largest was 65 feet by 7 feet 8 inches, surrounded by a dense thicket of suckers. Beech seedlings, however, seem to be much less common here than in Europe, and on moist ground are often suppressed by maple and other trees. The rate of growth of young trees in the Arboretum was about equal to that of the European beech at twenty years, and the bark of the latter was darker in colour. Near Ottawa Elwes gathered ripe fruit of the American beech [2]—which here is not a large or tall tree—in the end of September; the mast was smaller and less abundant than in the European beech, and the tree—as near Boston—did not seem to have the same tendency to outgrow and suppress other hardwoods which it shows in Europe. The roots, judging from seedlings sent from Meehan's nurseries at Philadelphia, are larger, deeper, and less fibrous than those of the European beech, though this may be caused by a deep soil. A good illustration of the American beech in the open is given in *Garden and Forest*, viii. 125, taken from a tree at South Hingham, Massachusetts.

The American beech is rare in collections in England. We have only seen specimens at Kew Gardens, Beauport, Tortworth, and Eastnor Castle. In no case do these attain more than 15 feet in height. As the tree, no doubt, was often planted even a century ago, and no large trees are known to exist in this country, it is very probable that, like many other species from the Eastern States, it will never reach timber size in this climate. The specimen from Eastnor Castle has very dull green leaves, somewhat cordate at the base, and probably belongs to the following variety.

Var. *caroliniana*, Loudon, *ex Lodd. Cat.* (1836).—In cultivation in Europe, distinguished from the common form by the leaves being more rounded at the base, said to be more dwarf in height, and to come out in leaf fifteen days before ordinary *Fagus ferruginea*.[3]

FAGUS ORIENTALIS. Caucasian Beech.

Fagus orientalis, Lipski, *Acta. Hort. Petrop.* xiv. 300 (1897).
Fagus sylvatica, Linnæus, β *macrophylla*, DC., and γ *asiatica*, DC. (*ex parte*), *Prod.* xvi. 2, 119 (1864).

Lipski says that the beech which occurs in the Caucasus, Asia Minor, and

[1] But Sargent says that it attains its largest size in the rich land of the Lower Ohio valley, and in the Southern Alleghanies, and that it often forms pure forests. He quotes an old author (Morton) as follows :—" Beech there is of two sortes, red and white, very excellent for trenchers or chaires, also for oares," and says that these different coloured woods, recognised by lumbermen, are produced by individual trees, which are otherwise apparently identical, and for which Michaux and Pursh tried to find botanical characters which he cannot allow to be specific.
[2] Sargent says that the sweet nuts are sold in Canada, and in some of the middle and western states.
[3] Jouin, " Les Hêtres " in *Le Jardin* (1899), p. 42.

North Persia, is a peculiar species.   Radde,[1] while not admitting it to be a distinct species, considers that it is a form which approaches the Japanese *Fagus Sieboldi*, Endl., rather than the typical European beech, which occurs in the Crimea.   Specimens in the Kew herbarium from the Caucasus, Paphlagonia, Phrygia, and Ghilan (a province of North Persia), differ markedly in fruit from the common beech.   This tree occurs throughout the whole of the Caucasus, both on the north and south sides, often ascending to the timber line, but descending in Talysch to the sea-level.   On the north side of the Caucasus the beech reaches to 5900 feet altitude ; while in the Schin valley, on the south side of the range, it attains 7920 feet.   It occurs mixed with other trees, or forms pure woods of considerable extent.   It sometimes occurs in the forests in the form of gigantic bushes (springing from one root), of which the individual stems measure 6 feet in girth, and are free from branches to 30 or 40 feet.   The largest trees recorded by Radde were :—one 380 years old, 7 feet in girth, and 123 feet high ; and another 250 years old, 8 feet 4 inches in girth, and 120 feet high, which contained 370 cubic feet of timber.

This species has been introduced into cultivation on the Continent, and is said[2] to have a crown of foliage more slender and more pyramidal than the common beech.

Fagus japonica.   Small Beech of Japan.   (Native name, *Inubuna*.)

Fagus japonica, Maximowicz, *Mél. Biol.* xii. 542 (1886).
Shirasawa, *Iconographie des Essences Forestières du Japon*, vol. i. t. 35, figs. 1-13 (1900).

This species is much rarer in Japan than *Fagus Sieboldi*, and was not seen by Elwes or Sargent, who says that it had not been collected since a collector in Maximowicz's employ found it on the Hakone mountains, and in the province of Nambu.   Very little is known about it, and it has not been introduced into Europe. Shirasawa, however, says it has the same distribution as *Fagus Sieboldi*, and grows almost always in mixture with it, but beginning at a lower level ; and that it often occurs in a bushy form, and does not attain the dimensions of the other species.

Fagus Sieboldi.   Common Beech of Japan.   (Native name, *Buna*.)

Fagus Sieboldi, Endlicher, *Gen. Suppl.* iv. 2, 29 (1847).
Fagus sylvatica, L., γ *asiatica*, DC. *Prod.* xvi. 2, 119 (1864).
Fagus sylvatica, L., δ *Sieboldi*, Maximowicz, *Mél. Biol.* xii. 543 (1886).
Shirasawa, *l.c.* t. 35, figs. 14-26.

This is the common beech which occurs in Japan, and it is considered by Japanese botanists[3] to be only a variety of the European beech.   Shirasawa[4] has given some details concerning its distribution, in connection with a figure which illustrates well the botanical characters of the species.   Sargent[5] was doubtful if the common beech in Japan was not quite identical in all respects with the European beech.

Elwes saw it in many places in Central Japan, but not in Hokkaido.   Near Nikko it grows to a large size at 2000-4000 feet, but not in pure woods, being, so

[1] Radde, *Pflanzenverbreitung in den Kaukasusländern*, 182 (1899).   [2] Schneider, *Laubholzkunde*, 152.
[3] Matsumura, *Shokubutsu-mei-i*, 123.   [4] Shirasawa, *l.c.* 86.   [5] Sargent, *Forest Flora of Japan*, 70.

far as he saw, always mixed with other trees, though Goto says[1] that it occurs in Honshu and in the southern half of Hokkaido in almost unmixed woods, and that in Aomori, Iwate, Echigo, and Yamagata, pure woods of vast dimensions are seen in the mountains above 1000 feet elevation. It is one of the most important trees for firewood and charcoal, but little valued for building. It grows well in shade, and continues to grow to a great age, sometimes attaining enormous size. The Ainos in old Japan are said to have used the tree for dug-out canoes. The largest trees measured by Elwes were in the Government forest of Atera, in the district of Kisogawa, where there were tall straight trees in mixed deciduous forests of beech, magnolia, oak, birch, and maple, about 100 feet high and 9-10 feet in girth. Here the wood was not of sufficient value to pay the expense of carriage.

FAGUS SINENSIS.

> Fagus sinensis, Oliver, in Hook. *Icon. Plant.* t. 1936 (1891); Diels, *Flora von Central China*, 284 (1901).
>
> Fagus sylvatica, L., var. *longipes*, Oliver, in scheda ad Hook. *Icon. Plant.* t. 1936 (1891); Franchet, *Jour. de Bot.* 1899, p. 90.
>
> Fagus longipetiolata, v. Seemen, in Engler, *Bot. Jahrb.* xxiii. *Beibl.* 57, p. 56 (1897).

This tree was discovered by Henry in the mountains south of the Yangtse, near Ichang, in Central China. It occurs scattered in deciduous forests at 3000-4000 feet altitude, and sometimes attains a considerable size, one tree being noted as 15 feet in girth. Von Rosthorn subsequently found the same species in the mountains south of Chungking, in Szechuan.

FAGUS ENGLERIANA.

> Fagus Engleriana, v. Seemen, in Diels, *Flora von Central China*, 285, cum figurâ (1901).
>
> Fagus sylvatica, L., var. *longipes*, Oliver, "var. *bracteolis involucri exterioribus spatulatim dilatatis*," Oliver, in scheda ad Hook, *Icon. Plant.* t. 1936 (1891).
>
> Fagus sylvatica, L., var. *chinensis*, Franchet, *Jour. de Bot.* 1899, p. 201.

This species was also discovered by Henry, but in the mountains north of the Yangtse from Ichang in Central China. Subsequently specimens were sent to Europe by Père Farges from North-East Szechuan, and by von Rosthorn from Southern Szechuan. It is a smaller tree than *F. sinensis*, and was seen by Henry on wooded cliffs.

Neither of the Chinese beeches form pure woods. A beech of considerable size was seen by Henry in Yunnan, in a mountain wood near Mengtse, at about 5000 feet elevation, and is possibly a distinct species. This rare tree is remarkable in that it extends the southern limit of the northern beeches to as low as 23° N.

---

[1] *Forestry of Japan* (1904), p. 22.

FAGUS SYLVATICA, COMMON BEECH

Fagus sylvatica, Linnæus, *Sp. Pl.* 998 (1753).   Loudon, *Arb. et Frut. Brit.* iii. 1950 (1838).

A large tree, commonly 100 feet high (attaining 130 to 140 feet under very favourable conditions), with a girth of 20 feet or more.   Bark[1] usually grey and smooth, but often in old trees becoming fissured and scaly, especially near the base.   Branchlets of two kinds; the short shoots ringed and bearing only a terminal bud in winter and one, two, or three leaves in summer; the long shoots slender, glabrous, with many leaves in two lateral rows (in winter the buds are seen arising from the upper side of the twig, the leaf-scars being on the lower side).

Leaves: deciduous, alternate, two-ranked, varying in size with altitude and vigour, those of trees at high elevations being much smaller; generally oval, somewhat acuminate at the apex, slightly unequal at the base, undulate or toothed in margin, with 6-10 pairs of lateral nerves, which with the midrib are raised on the under surface of the leaf, and are more or less pubescent.

Flowers: arising in the axils of the leaves of the young shoots; the male heads by long pendulous stalks, the female involucres by short erect stalks above the male flowers on the same branchlet or on separate branchlets.   The true fruits are usually two together enclosed in a woody involucre, which is beset by prickles.   Each fruit contains a seed, triangular in shape like the fruit containing it.   The seed hangs from the top of the cell and has no albumen.

Seedling: the seedling of the beech[2] has a long primary root and a stout radicle, 1-2 inches long, bearing 2 large sessile oval cotyledons, which are dark green above and whitish beneath.   The first true leaves of the beech are opposite, ovate, obtuse, and crenate, borne on the stem an inch or so above the cotyledons.   Above this pair other leaves are borne alternately, and the first season's growth terminates in a long pointed bud with brown imbricated scales.

The common beech is distinguishable at all seasons by its bark, which is only simulated by the hornbeam; but in the latter tree the stem is usually more or less fluted.   In winter the pointed buds, arranged distichously on the long shoots and composed of many imbricated scales, are characteristic; while in length they exceed

[1] There is much difference in the colour and roughness of the bark, which varies with age, soil, situation, and exposure. On the dry, sandy soil of Kew Gardens this bark of the beech is so different from that seen on calcareous soils that it might almost be mistaken for a hornbeam, and Elwes has observed the same in the Botanic Gardens at Edinburgh, where the trees are exposed to the salt east wind.   These variations are not, however, entirely caused by local conditions, but are sometimes found in trees standing close together.   Professor Balfour pointed out to Elwes two beeches in the Edinburgh garden of which one has the bark rough and scaly, and regularly comes into leaf fifteen to twenty days before another tree similar in size which grows next to it, whose bark is smooth and silvery.   Whether these variations are correlated with any differences in the wood does not seem to have been proved in England; but it is evident that for cold and exposed situations it would be advantageous to sow only the seed of the late leafing and flowering trees.

[2] The beech seedling has its cotyledons green and above ground; those of the oak and chestnut remain in the soil.   In the hornbeam, hazel, and alder, the cotyledons are aerial, but the first pair of true leaves above them are alternate.

those of any tree ordinarily cultivated in England, being about ¾ inch long. The buds of the European beech are wider at the middle than at either end ; while in the American beech they are as narrow in the middle as they are at the base.

VARIETIES

A great number of varieties of the common beech occur, some of which have originated wild in the forests, whilst others have been obtained in cultivation.

Var. *purpurea*, Aiton, Purple Beech. A complete account of the origin of this variety appeared in *Garden and Forest*,[1] 1894, p. 2. From this it would appear that a purple beech[2] discovered in the eighteenth century in the Hanleiter forest near Sondershausen in Thuringia, is the mother tree of those which now adorn the pleasure grounds of Europe and America. This is the only authenticated source from which horticulturists have derived their stock. The purple beech was, however, long known before the Thuringian tree was discovered. In Wagner's *Historia naturalis Helvetiæ curiosa* (Zurich, 1680) mention is made of a beech wood at Buch, on the Irchel mountain in Zurichgau (commonly called the Stammberg), which contains three beech trees with red leaves, which are nowhere else to be found. These three beeches are again referred to in Scheuzer's *Natural History of Switzerland*, published in 1706 ; and the legend is stated that according to popular belief five brothers murdered one another on the spot where the trees sprang up. Offspring of these trees were carried into a garden, where they still retained their purple colour. The purple beech has also been observed in a wild state in the forest of Darney in the Vosges.

The purple beech has delicate light red-coloured foliage, which is of a pale claret tint in the spring, becoming a deep purple in summer. In early autumn the leaves almost entirely lose their purple colour, and change to a dark dusky green. The buds, young shoots, and fruits are also purple in colour. The involucres are deep purple brown in autumn, becoming browner with the advance of the season. The purple beech often fails to fruit regularly ; still many individuals of this variety do produce fruit, and this has been sown, and in some cases produced plants almost all with purple leaves, not 5 per cent reverting to green.[3] The colour in the leaves, etc., is due to a colouring matter in the cells of the epidermis. The variety submits well to pruning or even to clipping with the shears ; and may therefore, if necessary, be confined within narrow limits or grown as a pyramid in the centre of a group of trees.

A fine purple beech[4] grows in Miss Sulivan's garden, Broom House, Fulham, which is 82 feet high and 12 feet 2 inches in girth.

---

[1] See also *Gartenflora*, 1893, p. 150.

[2] This tree is still living. See Lutze, *Mitth. des Thüringer Bot. Vereines*, 1892, ii. 28.

[3] Elwes saw at the Flottbeck Nurseries near Hamburg, formerly occupied by the celebrated nurseryman John Booth, a fine hedge of purple beech, which Herr Ansorge told him was raised from a cross between the purple and the fern-leaved beeches. Of the produce of this cross 20 to 30 per cent came purple, but none were fern-leaved. This coincides exactly with his own experience in raising from seed. But in *Mittheilungen Deutschen Dendrologischen Gesellschaft*, 1904, p. 198, Graf von Schwerin describes as *F. sylvatica ansorgei* a hybrid from these two varieties which seems to combine the characters of both.

[4] Figured in *Gard. Chron.* 1898, xxiv. 305. See also *ibid.* 1903, xxxiii. 397, for notes on sub-varieties of the purple beech.

Another occurs at Hardwick, Bury St. Edmunds, the seat of G. M. Gibson Cullum, Esq., which in 1904 was 11 feet 9 inches in girth, and about 80 to 90 feet in height. Bunbury[1] considered this to be the finest purple beech in England, and says it produces abundance of fruit, from which young trees have been raised.

Var. *cuprea*, Loddiges, Copper Beech.—This is only a sub-variety of the purple beech, distinguished by its young shoots and leaves being of a paler colour. The largest purple or copper beech which Elwes has seen is in the park at Dunkeld, Perthshire, not far from the Cathedral. This measures 86 feet high, with a girth of 15 feet 3 inches, and does not show any evidence of having been grafted. There is a very fine one at Corsham Court, the seat of General Lord Methuen, 85 to 90 feet high, by 14 in girth, forking at about 10 feet. At Scampston Hall, Yorkshire, Mr. Meade-Waldo tells us of two large spreading trees on their own roots, 11 feet 6 inches and 10 feet 6 inches in girth respectively. At Beauport, Sussex, the seat of Sir Archibald Lamb, Bart., a copper beech measured $12\frac{1}{2}$ feet in girth in 1904. At Syston Park, Lincolnshire, the seat of Sir John Thorold, Bart., there is one nearly as large (12 feet 2 inches girth). A copper beech at Bell Hall, York, which was planted in 1800, measured in 1894, 9 feet in girth, the diameter of the spread of the branches being 74 feet. At Castle MacGarrett, Claremorris, Ireland, the seat of Lord Oranmore, there is a beautiful copper beech, which in 1904 was 70 feet high and 9 feet 10 inches in girth. In Over Wallop Rectory grounds, in Hampshire, a copper beech measured 9 feet 4 inches in 1880.

Two fine trees occur at Clonbrock, in Co. Galway, the seat of Lord Clonbrock. One measured in 1904 a length of 76 feet and a girth of 12 feet 9 inches. The other was 7 feet 6 inches girth in 1871, and in 1880 it had increased to 8 feet 5 inches.

The copper beech[2] is rarely used as a hedge, but there is one in the gardens of Ashwellthorpe Hall, Norwich, which is 138 yards long, 8 feet high, and about 5 feet through. It was planted about seventy years ago from seedlings by the Hon. and Rev. R. Wilson. The colouring in spring is very beautiful.

There is a sub-variety[3] of the copper beech in which the leaf is edged with pink whilst young, but later in summer it becomes nearly like the type. This variety has been called *Fagus purpurea roseo-marginata*, and it has been recommended as a hedge-plant, to be clipped two or three times during summer so as to obtain several crops of young shoots.

Var. *atropurpurea*.—The leaves in this are of a darker colour than in the ordinary purple beech.

Var. *atropurpurea Rohani* is quite different from the last, as the form of the leaves is similar to that of the fern-leaved beech, but their colour is like that of the copper beech.

Var. *purpurea pendula*.—This is a weeping form of the purple beech. It is of slow growth.

[1] *Arboretum Notes*, p. 117.   [2] *Garden*, July 30, 1904, Answers to Correspondents.
[3] *Gard. Chron.* June 23, 1888, p. 779.

Var. *Zlatia*, Späth,[1] Golden Beech.—This was found wild in the mountains of Servia by Professor Dragaschevitch. It is known in Servia as *Zladna bukwa* (golden beech).

Var. *striata*, Bose.[1]—This was discovered many years ago in a forest in Hesse. Soon after opening, the leaves show a regular golden striation parallel with the nerves, and this appearance lasts till the leaves fall off in autumn. It was introduced in 1892 by Dippel.

Various other coloured varieties have been obtained by horticulturists. In var. *variegata* the leaves are particoloured with white and yellow, interspersed with some streaks of red and purple. In var. *tricolor* the leaves are dark purplish green, spotted with bright pink and shaded with white. There are also gold-striped (var. *aureo-variegata*) and silver-striped (var. *argenteo-variegata*) varieties.

Var. *heterophylla*, Loudon, Fern-leaved Beech.—The leaves are variously cut, either in narrow shreds like some ferns, or in broader divisions like the leaves of a willow. This variety has received a great number of names, as *laciniata*, *comptoniæfolia*, *incisa*, *salicifolia*, *asplenifolia*, etc. The tree occasionally bears normal and cut leaves on the same twig, or normal and cut leaves on different twigs. It bears fruit occasionally, which, according to Bunbury,[2] is smaller than that of the common beech, the cupule being shorter in proportion to the nuts. The leaf-buds are considerably smaller than those of the common form; and the twigs are often very pubescent. The origin of this variety is unknown.

There is a good specimen of this tree at Devonshurst House, Chiswick, which measured in 1903 55 feet in height, and 8 feet 2 inches in girth at 3 feet, just under a great horizontal branch.

At Barton, Bury St. Edmunds, a fern-leaved beech in 1904 was 53 feet high, with a girth of 5 feet 1 inch. This tree[2] was planted in 1831, but grew slowly, in 1869 being only 15 feet high, with a trunk 3 feet round. In 1868 the tree bore some twigs with ordinary leaves; and it first fruited in 1869, the crop being a very small one.

There are large and well-shaped trees of this form at Strathfieldsaye measuring 50 feet by 7 feet 5 inches; at Fawley Court of the same size exactly, and weeping to the ground; and at Stowe near Buckingham.

Var. *quercoides*, Pers., Oak-leaved Beech.—The leaves in this variety are long-stalked, with an acute base and acuminate apex; margins pinnately and deeply cut, the individual segments being acute.

Var. *cristata*, Lodd. (also known as var. *crispa*).—Small and nearly sessile leaves, crowded into dense tufts, which occur at intervals on the branches. This form rarely attains a large size.

Var. *macrophylla* (also known as *latifolia*).—The leaves in this form are very large. In a specimen at Kew, from the garden of the Horticultural School at Vilvorde, they attain 7 inches in length and 5 inches in width. A large specimen of this tree, some fifty years old, occurs at Enys in Cornwall. The buds, as might

[1] *Gard. Chron.* 1892, xii. 669. This is an account of Späth's novelties by Dr. Edmund Göze of the Greifswald Botanic Gardens. [2] *Arboretum Notes*, p. 118.

I

C

be expected, in this variety are considerably larger than those of the ordinary form.

Var. *rotundifolia*, Round-leaved Beech.[1]—The leaves are very small, round, and bright green, and are set close on the twigs. This variety has an upright habit of growth, and was introduced in 1894 by Jackman of Woking.

Var. *grandidentata*.—A form with conspicuously toothed leaves.

Var. *pendula*, Loddiges, Weeping Beech.—Several forms of this variety occur, but in all the smaller branches hang down. The main branches are irregularly disposed, so that the tree often has a very rugged outline. This variety should be grafted at a good height, as otherwise many of the pendulous branches will lie upon the ground; and the main branches, if they show a tendency to droop too much, should be supported. Weeping beeches may be tall and slender, or low and broad, or quite irregular, depending upon the direction of the larger branches, which may grow outwards or upwards, or in almost any direction; the smaller branches only are uniformly pendulous.

The weeping beech has been observed wild in the forest of Brotonne, in Seine-Inférieure, France.

A good example of a tall, slender, weeping beech may be seen near Wimbledon Common, on the estate lately owned by Sir W. Peek. A fine specimen occurs at Barton, which in 1904 was 77 feet high and 5 feet 2 inches in girth. Elwes has noted a very picturesque and well-shaped one at Endsleigh, near Tavistock, the Devonshire seat of the Duke of Bedford. Several have been figured in the *Gardeners' Chronicle*, e.g. a group of three trees[2] at Ashwick Hall, Gloucestershire, which were planted about 1860. In the Knap Hill Nursery[3] at Woking, and in the nursery[4] of R. Smith and Co. at Worcester, there are fine specimens. Another good specimen,[5] occurring in Dickson's nursery at Chester, is figured in the *Garden*.

Many forms of weeping beech have been described as sub-varieties, as *purpurea pendula*, mentioned above; var. *miltonensis*, with branches less pendulous, found wild in Milton Park, Northamptonshire; var. *borneyensis*, found wild in the forest of Borney, near Metz, and described as having an erect stem and distinctly pendulous branches; var. *pagnyensis*, discovered in the forest of Pagny in the department of Meurthe-et-Moselle in France; var. *remillyensis*, found in the forest of Remilly, near Metz.

Var. *tortuosa*, Parasol Beech.[6]—In this curious form, the branches, both large and small, and the branchlets are all directed towards the ground. It is not to be confounded with the preceding variety, in which only the slender branches are pendulous; and is analogous rather to the weeping ash. Beeches of this form have, even in old age, a very short and twisted stem, with a hemispherical crown, which sometimes touches the ground; and it scarcely ever grows higher than 10 feet. This variety has been found wild in France, in the forest of Verzy, near Rheims, and also

[1] *Gard. Mag.* 1894, p. 339, with figure.   [2] *Gard. Chron.* June 20, 1903, fig. 155.
[3] *Ibid.* Dec. 24, 1870, p. 70.   [4] *Ibid.* Dec. 29, 1900, suppl.   [5] *Garden*, Dec. 5, 1903, p. 167.
[6] For a complete account of the occurrence of this curious form in the forests of the east of France, see Godron, *Les Hêtres tortillards des environs de Nancy*, Mém. de l'Acad. de Stanislas, Nancy, 1869. Godron says that their growth is infinitely slower than that of normal beech. See also *Rev. Hort.*, 1861, p. 84, and 1864, p. 127.

in the neighbourhood of Nancy. Fruits of this form have been sown in the garden of the Forest School of Nancy, and have reproduced the twisted form in about the proportion of three-fifths; the other two-fifths of the fruit produced form like the common beech and intermediate varieties.[1]

Many other varieties have been described; and other forms possibly occur wild which have not been noticed. Major M'Nair sent to Kew in 1872 from Brookwood, Knaphill, Surrey, a specimen from a tree growing there, and reported to be in vigorous health, in which the leaves are remarkably small and have only four pairs of lateral nerves. (A. H.)

DISTRIBUTION

The beech is indigenous to England. Remains of it have been found in neolithic deposits at Southampton docks, Crossness in Essex, in Fenland, in preglacial deposits in the Cromer forest bed, and at Happisburgh, Norfolk.[2] Names of places of Saxon origin, in which the word beech occurs are very common, as Buckingham, Buxton, Boxstead, Boxford, Bickleigh, Boking, etc. The existence of the beech in Britain in ancient times has been questioned on account of the statement by Julius Cæsar[3] that Fagus did not occur in England. H. J. Long[4] has discussed what tree the Romans meant by Fagus, and the evidence is conflicting. Pliny[5] described as Fagus a tree which is plainly the common beech. However, Virgil's[6] statement that *Castanea* by grafting would produce *fagos* indicates rather that Fagus was a name used for the sweet chestnut; and this view is confirmed by the fact that out of the wood of Fagus the Romans made vine-props and wine-casks. The Latin word Fagus is derived immediately from the Greek φηγός; and the φηγός of Theophrastus is certainly the chestnut, probably the wild tree which is indigenous to the mountains of Greece. Cæsar's statement probably implies that in his day the sweet chestnut did not occur in Britain.

The beech is not believed to be indigenous in Scotland and Ireland,[7] and no evidence is forthcoming of its occurrence in prehistoric deposits in those countries. An able writer in *Woods and Forests* (1884, June 11, p. 404) contests this view, and speaks of the existence of two beech woods in the north of Scotland, not 10 miles from the most easterly point of Britain, where the trees were larger than any other timber tree, not excepting the Scotch fir, and where it produced fertile seed, while that of the oak was abortive. These woods were high and exposed, but the soil was good. In view of the way in which the beech ascends in the Vosges and the Jura to cold, bleak situations, finally becoming at 4000 feet a dwarf shrub, which

---

[1] The parasol beech, or a form closely like it, has been found in Ireland, according to a correspondent of *Woods and Forests*, Jan. 1885, who writes as follows :—"Near to Parkanour, in Tyrone, the residence of Mr. J. Burgess, stand two beeches, which at a short distance resemble heaps of leaves more than trees. They were found in the woods sixty years since, and are from 6 feet to 8 feet in height and 15 feet diameter, and of dense drooping habit. Upon creeping inside, I found them to branch off at 2 feet or 3 feet from the ground, where one was nearly 5 feet in circumference. The arms and branches are not unlike corkscrews. The inferior branches and matted rubbish, if cleared out, would greatly improve their appearance, as the singular growth would then be visible. They might, if sent out, become a valuable adjunct to the upright yew, which flourishes in Ireland, the finest of which I have yet seen being 24 feet high and 12 feet through, and well filled in the centre.—C. I."

[2] C. Reid, *Origin of British Flora*, 28, 69, 146.   [3] *B. G.* v. 12.
[4] Loudon, *Gard. Mag.* 1839, p. 9.   [5] *N. H.* xvi. 7.   [6] *Georg.* ii. 71.
[7] The name in Irish is *crann sleamhain*, the "slippery tree," so-called from the smoothness of the bark.

forms the timber line, it would be remarkable if the beech had not in early days gained a footing in Scotland and Ireland. The mere negative evidence is of little value, as scarcely any scientific work has yet been done in the way of exploration of the peat-mosses and other recent deposits ; and the woods, from which are made the handles of numerous prehistoric implements preserved in our museums, have rarely been examined.[1]

The beech occurs in a wild state throughout the greater part of Northern, Central, and Western Europe, usually growing gregariously in forests which, when undisturbed by man, have a tendency to spread and take the place of oak, which, owing to its inability to support such dense shade, is often suppressed by the beech.

In Norway, according to Schubeler,[2] it is called bok, and is wild only near Laurvik, where he believes it to be truly indigenous, and is a small tree, the largest he measured being 7 feet 4 inches in girth. At Hosanger, however, a planted beech had in 1864 attained 75 feet at 81 years old, with a diameter of 27 inches. It ripens seed as far north as Trondhjem in good years, and exists in Nordland as far north as lat. 67°.56.

In Sweden its most northerly wild habitat is Elfkalven, lat. 60°.35, though it has been planted as far north as lat. 64°.

In Russia the beech extends only a little way,—its eastern limit in Europe passing the Prussian coast of the Baltic between Elbing and Königsberg, about 54° 30′ N. lat., and running south from Königsberg, where the last spontaneous beeches occur on the Brandenburg estate, continuing through Lithuania, eastern Poland, Volhynia, where beech woods still occur between lat. 52° and 50°, and Podolia to Bessarabia. It is absent from the governments of Kief and Kherson, but re-appears in the Crimea, where, however, it is only met with in the mountains of the south-east coast. In the Caucasus, Persia, and Asia Minor it is replaced by the closely allied species, *Fagus orientalis*.

In Finland and at St. Petersburg it exists as a bush only, but is not wild. On the southern shores of the Baltic it forms large forests, and in Denmark is one of the most abundant and valuable timber trees, growing to as large a size and forming as clean trunks as it does farther south. Lyell speaks of it as follows :[3]—" In the time of the Romans the Danish isles were covered as now with magnificent beech forests. Nowhere in the world does this tree flourish more luxuriantly than in Denmark, and eighteen centuries seem to have done little or nothing towards modifying the character of the forest vegetation. Yet in the antecedent bronze period there were no beech trees, or at most but a few stragglers, the country being then covered with oak."

At page 415 he says further—" In Denmark great changes were taking place in the vegetation. The pine, or Scotch fir, buried in the oldest peat, gave place at length to the oak ; and the oak, after flourishing for ages, yielded in its turn to the

---

[1] In a paper by H. B. Watt on the "Scottish Forests in Early Historic Times," printed in *Annals of the Andersonian Nat. Soc.* ii. 91, Glasg., 1900, which contains many interesting particulars of the oak and other trees, no mention is made of the Beech. In the Highland Society's *Gaelic Dictionary* (1828), *faidhbhile* is given as the word for beech ; here *faidh* is cognate with *fagus*, *bhile* being one of the Gaelic terms for *tree*. This name is also known in Ulster.

[2] Schubeler, *Viridarium Norvegicum*, vol. i. 521.   [3] Lyell, *Antiquity of Man*, 2nd. ed. 1873, pp. 17, 415.

beech ; the periods when these three forest trees predominated in succession tallying pretty nearly with the ages of stone, bronze, and iron in Denmark."

All over Germany, except in the sandy plains of the north, it is one of the principal forest trees ; but, so far as we have seen, does not—or is rarely allowed to attain—such a great size as in England. In Central and Southern Germany and in Eastern and Southern France it seems to be indigenous only in hilly districts and mountains.

In the north of France it attains perfection, and forms very large forests, usually mixed with oak, which sometimes contain trees of immense height, but is not planted as an ornament to parks as much as in England.

According to Huffel's *Économie Forestière*, 362 (1904), the finest beech forest in France is that of Retz, also called Villers Cotterets, which contains 37,000 acres, on a soil composed of deep sand, mixed with a slight proportion of clay. The trees consist almost entirely of beeches, there being only a small number of oaks and hornbeams. In the best plot of this forest, the canton of Dayancourt, which is 30 acres in extent, there were, in 1895, 1998 beech trees, 20 oaks, and 16 hornbeams. The beeches contain 329,433 cubic feet of timber, and reach a height of nearly 150 feet with clean stems of 80 to 90 feet. Their age in 1895 was 183 years, and they were considered to have reached their maximum development and to be on the point of going back.

In an account of the beech, Mr. Robinson has stated in *Flora and Sylva* that in the forest of Lyons-la-Forêt, near Rouen, beeches of 160 feet in height are found ; but on asking my friend M. Leon Pardé, inspector of forests at Beauvais (Oise), near Paris, whether this statement could be confirmed, he was good enough to send me a letter from the forest officer there, who says that the tallest beech known in France is the one which I saw in the Forêt de Retz, when the English Arboricultural Society visited France in 1903,[1] the height of which was given as 45 metres, about 147 feet. This tree measured 13 feet 2 inches in girth, and was straight and clear of branches to 91 feet. It was estimated by the English measurement to contain 560 cubic feet to the first branch, or 700 feet in all. This letter goes on to say that the tallest trees at Lyons-le-Forêt do not, in his opinion, exceed a total height of 35 metres, though one has doubtfully been stated to attain $37\frac{1}{2}$ metres.

Two of the finest and tallest beeches in France are the one called "La Bourdigalle" in the Forêt de Lyons at La Haye (Seine Inf.), which is 35.80 metres high by 5.55 metres at 1 metre, and is supposed to be from 375 to 575 years old.[2]

Another called "Le Trois Hêtres," in the forest of Brotonne at Guerbaville (Seine Inf.), has three straight clean stems rising from a single base to a height of nearly 35 metres, with a girth at 1 metre of about 18 feet. This very remarkable tree is figured on plate xi. of the work cited below.[2]

In Switzerland pure beech forest is found as high as 4500 feet, and at 5000 assumes a shrubby habit.

In the Austrian Alps and Carpathians it is also a common tree, forming vast forests, which are sometimes pure, sometimes mixed with other trees.

---

[1] *Trans. Eng. Arb. Soc.* v. pt. ii. p. 209.   [2] Gadeau de Kerville, *Les Vieux Arbres de la Normandie*, 143 (1893).

In Italy it is found only in the mountains; in the Apennines it is one of the dominant trees at from 3000 to 5000 feet. In the Sila mountains of Calabria, Elwes found it covering the mountains above the limit of chestnut, at from 3000 to 5000 feet and upwards. It is usually coppiced for charcoal and firewood; but it attains a considerable size, the largest measured being about 90 feet by 10-12 in girth. Here it is often mixed with the Calabrian pine. In Sicily it finds its southern limit on Mount Etna, where it ascends to 7200 feet.

In Spain the beech occurs in the Pyrenees and in the northern provinces only, its most southerly known habitat being in lat 40° 10′ east of Cuença. In Portugal it has not, so far as we know, been recorded to exist.

The finest natural beech forests seen by us in Europe are on the northern slopes of the Balkans, where it grows as pure forest from near the foot of the mountains up to about 4000 feet. The trees are very straight and clean, but are being rapidly felled in those places where they are most accessible. Boissier[1] says that the beech occurs in northern Greece on Mounts Pindus and Pelion. Elwes found it in Macedonia, on the north side of Mount Olympus.[2]

(A. H. and H. J. E.)

CULTIVATION

Seed is without doubt the best means of reproducing the tree, and I am inclined to think that the best and cleanest trunks are produced by seedlings which have never been transplanted, but opinions differ on this question. Seed is only produced in quantity at intervals of several years, and in some years a large proportion of the seeds, even in districts where the beech grows well, are mere empty husks.

The season of 1890 was probably the best for beech-mast in England which had occurred for many years, and I took particular pains, by enclosing certain spots where I found a number of germinating seeds in the following April, to protect them. But a severe frost, which occurred in the middle of May, destroyed all or nearly all the seedlings in the open, and those whose germination had been delayed by dense shade, or a thick covering of leaves, mostly withered away in the dry summer which ensued, before their rootlets had become established in the ground. Notwithstanding this, in most woods where rabbits, pheasants, and wood-pigeons are not so abundant as to devour all the seedlings and seeds, a good number of seedling beech of the year 1901 may still be found, and in the New Forest and elsewhere the ground in suitable spots is covered with seedlings.

Whether the seed should be sown when ripe or kept until the following spring is a question which must be decided by local conditions and experience, but where the danger of late spring frosts is great, I should prefer keeping it in an airy, dry loft spread thinly on a floor until April, or even the first week in May, as if February and March are mild, it will germinate in March and run great risk of being frozen in April or May. On March 11, 1901, I found a quantity of

[1] *Flora Orientalis*, iv. 1175.
[2] Halácsy, *Consp. Flor. Græcæ*, iii. 124 (1904), says that the beech forms in Greece large woods in the mountains, and gives its distribution as follows:—Thessaly—Mountains of Pindus, Chassia, Olympus, Ossa, and Pelion; Acarnania—Mount Kravara; Ætolia—Mount Oxyes.

beech-mast on the lawn at Heythrop Park which had already germinated and had the radicle protruding as much as $\frac{1}{2}$ inch. I gathered a basketful and sowed it two days later, covering the drills with beech-leaf mould. Most of this was above ground in April, and where not protected by branches over the beds, was destroyed by frost. Stored seed sown at the same time was almost all devoured by mice and rooks, which seemed to follow the drills with great care, whilst seed sown broadcast on a freshly ploughed surface and covered by one turn of a harrow, produced a certain number, but still a very small proportion of plants. These were, in June 1904, still very small and stunted, not more than 3-5 inches high, whilst seedlings of the same age raised on good rich sandy soil in an Edinburgh nursery were from $1\frac{1}{2}$ to 2 feet high.

In the autumn of 1902 I found it impossible to procure any beech-mast in Great Britain, and after many inquiries procured some German seed early in April. Part of this was dibbled in a field of wheat, but so few plants could be found when the wheat was cut that the experiment was a practical failure. I sowed a part of this seed early in May in the garden, which germinated in June, and thus escaping spring frosts it grew without a check, and the seedlings were 4 to 6 inches high in the autumn.

Judging from these results it appears to me that, except in woods or where there is shelter, it is not economically desirable to raise beech from seed where it is to grow, and that spring sowing is preferable to autumn.

Seedlings are easy to transplant if their roots are not allowed to become dry, and the percentage of loss in 20,000 sent to me from Edinburgh in the winter 1902-3 was not more than 5 to 10 per cent. But if the trees are older and the roots are bad or have been heated in transit, or exposed too long to the air, the loss will be very great; and in most cases I should not plant out on a large scale trees of over two years old two years transplanted, though for specimen or lawn trees they may be safely moved when 6 to 10 feet high, or even more, if properly transplanted every two years.

SOIL AND SITUATION

Though the beech will grow on almost any soil except pure peat and heavy wet clay, it comes to its greatest size and perfection on calcareous soil or on deep sandy loam, and usually in pure woods unmixed with other trees.

The finest beech woods in England are, or rather were, in the Chiltern Hills, Bucks, in the neighbouring counties of Oxford and Herts; in the valleys of the Cotswold Hills; and in Sussex.

Sir John Dorington, M.P., tells me that he cut 2 acres 1 rood 13 poles of beech on a steep bank opposite his house at Lypiatt Park, Stroud, in 1897, growing on thin oolite limestone brash, which at 1s. 2d. per foot produced £562, equal to about 9634 feet. And off 4 acres of the same wood in 1875 he sold beech to the value of £1100, being at the rate of £275 per acre. This was supposed to be about 150 years old, and is the best actual return of value from timber on such land which I know of. He also bought a beech wood of 26 acres growing on similar soil in 1898, on which the timber, supposed to be about seventy years old, was valued at £2200, equal to £85 per acre. He cut £600 worth of thinnings out of it the year following ; and as the trees are growing fast, considers that it might now be valued at the same price per acre.   Sir John considers, from experience in his own plantations, that planted beech will do as well as when naturally seeded.   His old woodman, now dead, was for long of a contrary opinion, but changed his mind latterly from his own experience.

It is necessary to say something about the actual conditions and returns from the Buckinghamshire beech woods, which have been held up by some writers as an example of what may be done by following the system known as *jardinage* in France, which consists in thinning out the saleable trees every ten or twelve years and allowing natural seedlings to come up in their places.

During a visit of the Scottish Arboricultural Society on July 30, 1903, to this district, in which I took part, it was stated by one of the principal land agents in the district that £2 per acre was a common return over an average of years on woods managed on this system, which seems to have grown up during the last sixty years, partly through the legal disability of the owners to make clear fellings, and partly owing to the regular demand for clean beechwood of moderate size for chair-making. But what I saw myself led me to believe that though such a return may have been obtained for a short period on the best class of beech woods, it is not likely to continue, and that if an owner had a free hand and was not liable for waste, clear felling of the mature timber about once in 60-100 years would probably in the long-run be a better system.   And this opinion was confirmed by Mr. George James, agent for the Hampden estate, who thinks that 15s. per acre, which is about the average rateable value of these woods, is as much as they are actually worth, and that when you get fine timber clean and well grown, as on Mr. Drake's estate at Amersham, many natural seedlings do not occur, but that on Earl Howe's estate at Dunn where, forty years ago, all or nearly all of the timber was cut, there is a good growth of young seedlings.

Professor Fisher of Cooper's Hill has written a very instructive article[1] on the Chiltern Hill beech woods, in which he states that these are probably the northern and western British limit of the indigenous beech forest, which was probably eradicated during the glacial period in the north of England ; though remains found in the submarine forest-bed at Cromer, in Norfolk, prove that it existed before this period farther east.   He quotes measurements taken by Mr. A. S. Hobart Hampden, now director of the Forest School at Dehra Dun, India, which show that on the average it takes ninety years in this district for beech to attain 3 feet in girth at breast height, and that a full crop of seed cannot be expected from trees much younger than eighty years when grown in dense order.   He agrees with me that in many of the woods, including those which belong to Eton College, over-thinning has been prevalent, and states that rabbits and brambles have in many cases prevented the natural regeneration from being as complete as it must be to keep such woods in profitable condition under the decennial selection system.[2]   And as

---

[1] *Land Agent's Record*, April 9 and 16, 1904.
[2] A paper by Mr. L. S. Wood, in the *Trans. Eng. Arbor. Soc.* v. 285 (1903), gives many particulars of the beech woods in this district.

the furniture factories of High Wycombe are now largely supplied with American birch and other foreign timber, which can be imported at a cheaper rate than beech is locally worth, I am inclined to think that where these woods have become too thin to be profitable, they would pay better if the seeding of ash—which grows well on this land though not to the largest size—was encouraged, and the vacant spaces filled up with larch, which, when mixed with beech, usually keeps healthy and grows to a larger size than it does alone.

It is probable, however, that as our coal supplies diminish, the value of firewood in England will increase, and as beech is one of the best firewoods we have, and one of the most economical to convert into suitable sizes, I should advise its being more largely planted in districts where coal is distant and costly.

As a nurse to other forest trees, especially larch and oak, it has a value greater than any deciduous tree, because, if not allowed to overtop its neighbours, its shade and the decay of its leaves preserve the soil in a cool, moist, and fertile condition. On poor calcareous and chalk soil it is specially valuable, and should be planted in mixture with most kinds of other trees, provided rabbits can be permanently excluded; but on account of its thin bark it is never safe in a deep snow or in hard winters from rabbits, which will bark the roots of trees 100 years old as readily as young trees.

The distance apart at which beech should be left in plantations, must depend on the goodness of the soil and on the size at which the trees can be most profitably cut. The better the land the thicker it may stand, but on really poor soil it grows so slowly if crowded, that as soon as it has attained a sufficient height and cleaned itself from branches up to 30-50 feet, it should be thinned to about 150 trees or even less to the acre. And I have often observed that on soils which are not naturally favourable for beech, it will not under any circumstances grow so straight and clean as in woods where natural regeneration is easy.

Notwithstanding what Loudon and some German foresters say about the beech being unfit for coppice-wood, I can show beech stools of considerable age which have been regularly cut over at intervals of about eighteen years for at least a century; whilst the growth of shoots from the stool on the dry rocky bank in Chatcombe Wood, near Seven Springs, on the Cotswold Hills, is faster than that of ash similarly treated. In the mountains of Calabria also, I have seen hillsides covered with beech scrub which appeared to have been coppiced for firewood for a very long period. Therefore, in cases where the beech has been planted merely as a nurse to oak or other trees, and there is no deciduous tree better adapted to this purpose, I should not hesitate to cut over the trees if they seemed likely to smother their neighbours, with the expectation of getting a quantity of excellent firewood or small poles fit for turning, fifteen to twenty years later.

As a clipped hedge the beech is useful, but does not grow so fast at first as the hornbeam. An excellent example of this fact may be seen near the entrance to Dr. Watney's place at Buckholt, near Pangbourne, where the two are growing in the same hedge; the beech treated in this way keeps its leaves all the winter and makes good shelter.[1]

[1] Cf. Loudon, *loc. cit.* p. 1965.

I

D

Beech Avenues

Sir Hugh Beevor has sent me a photograph of a remarkable avenue of beech trees called Finch's Avenue, near Watford, which is composed of straight, clean, closely planted trees up to 120 feet high (Plate 2).

As an avenue tree the beech is one of the most stately and imposing that we have; but probably because of the difficulty of getting tall, straight standards from nurseries, and their tendency to branch too near the ground when planted thinly, they are not so much in vogue as they were two centuries ago. One of the finest examples I know of in England is the grand avenue in Savernake Forest, the property of the Marquess of Ailesbury. This was planted in 1723, and extends for nearly 5 miles from Savernake House to the hill above Marlborough. It is described and figured in the *Transactions of the English Arboricultural Society*, v. p. 405, and though the trees are not individually of quite such fine growth as those at Ashridge, yet, forming a continuous green aisle meeting overhead, for such an immense distance, it is even more beautiful than the elm avenue at Windsor, or the lime avenue at Burghley, and surpasses both of them in length. The Savernake avenue, however, is not like those above mentioned, planted at regular distances, but seems to have been cut out of a belt.

The beech avenue at Cornbury Park, the property of Vernon Watney, Esq., to whom I am indebted for the following particulars, is, on account of the great size of the trees, one of the most imposing in England. It was probably planted or designed by John Evelyn, whose diary, 17th October, 1664, says: "I went with Lord Visct. Cornbury to Cornbury in Oxfordshire, to assist him in the planting of the park, and beare him company, dined at Uxbridge, lay at Wicckam (Wycombe)." They reached Cornbury the following day, and among the entries for that day is the following: "We designed an handsom chapell that was yet wanting as Mr. May had the stables, which indeed are very faire having set out the walkes in the park and gardens." This Lord Cornbury who, after his father's death, became Lord Clarendon, records in his diary, "1689, September 25. Wednesday.—The elms in the park were begun to be pruned." This avenue is 800 yards long, and runs from the valley where the great beech grew, up the hill to the house. Many of the trees seem to have been pollarded when young at about 15 feet high, but have shot up immense straight limbs to a height of 100 to 110 feet, some even taller.

The Ten Rides in Cirencester Park affords a good illustration of the value of the beech for bordering the broad rides through a great mass of woodland; but the trees here, as at Cornbury and in so many of our old parks, have seen their best days, and when blanks are made by wind or decay, it is beyond the power of man to restore the regular appearance of such a vista.

Whatever pains may be taken to replant the gaps, the trees never seem to run up as they do when all planted together, and the art of planting avenues does not seem to be so well understood or so much practised now, as it was in the seventeenth and eighteenth centuries.

Remarkable Trees

As an instance of the rapid growth of the beech, I will quote from a letter of Robert Marsham of Stratton Strawless, near Norwich, to Gilbert White, dated 24th July 1790, in which he says : "I wish I had begun planting with beeches (my favourite trees as well as yours), and I might have seen large trees of my own raising. But I did not begin beeches till 1741, and then by seed ; and my largest is now at 5 feet, 6′ 3″ round, and spreads a circle of + 20 yards diamr. But this has been digged round and washed, etc." In Gilbert White's reply to this letter, dated Selborne, 13th August 1790, he says: "I speak from long observation when I assert, that beechen groves to a warm aspect grow one-third faster than those that face to the N. and N.E., and the bark is much more clean and smooth."

Marsham, replying to White on 31st August (it seems to have been at least fifteen days' post in those days from Norfolk to Hants), says : "Mr. Drake has a charming grove of beech in Buckinghamshire, where the handsomest tree (as I am informed by a friend to be depended on) runs 75 feet clear, and then about 35 feet more in the head. I went on purpose to see it. It is only 6 F. 6 I. round, but straight as possible. Some beeches in my late worthy friend Mr. Naylor's park at Hurst-monceux in Sussex ran taller and much larger, but none so handsome." In a later letter he speaks of one being felled here in 1750 which "ran 81 feet before it headed."

Sir Hugh Beevor informs me that he found it impossible to identify with certainty the trees measured at Stratton Strawless by Marsham, which we shall have occasion to allude to later.[1]

It would be impossible to mention more than a few of the finest beech trees in this country, but the photographs which have been reproduced represent a few of those which I have seen myself.

In Hants there are many fine beeches in the New Forest, of which the wood called Mark Ash contains some of the most picturesque, and is to my eyes one of the most beautiful woods from a naturalist's point of view in England, or even in Europe, though it is, like so many of the fine old woods in the New Forest, deteriorating from causes which are described elsewhere. One of the finest trees here is over 100 feet high and 24 feet in girth, dividing at about 10 feet into six immense erect limbs, and entirely surrounded, as are many of the trees in this wood, by a dense thicket of holly.

There is another beech in Woodfidley in the New Forest which Mr. Lascelles considers the finest beech in the forest, and of which the measurement as given by him is 120 feet high, 14 feet 6 inches in girth at 5 feet, carrying its girth well up, with an estimated cubic content of 650 feet.

In Old Burley enclosure is another magnificent beech, rather shut in by other trees, and therefore difficult to measure for height. I estimated it at 110 feet high. The girth was 18 feet, dividing at about 25 feet into two main trunks, which carried a

---

[1] Cf. *Trans. of the Norfolk and Norwich Nat. Soc.* ii. 133-195.

girth of perhaps 8 feet up to a great height. I have no doubt this tree contains 700 to 800 feet of timber.

At Knole Park, near Sevenoaks, there are some splendid trees of the park type, with very wide-spreading limbs, two of which are known as the King and Queen Beeches. The King Beech is surrounded by a fence, and many of its branches are supported by chains. Strutt, who figures it, gives its height as 105 feet by 24 in girth at 13 feet. When I measured it in 1905 it was about 100 feet by 30 in girth at 5 feet, with a bole 10 feet high. It has the largest girth of any beech I know of now standing in England (Plate 12). The Queen Beech is 90 to 100 feet high and 28 feet in girth. I am not sure whether this or the last is the one recorded by Loudon, iii. 1977, as having a diameter of 8 feet 4 inches, a height of 85 feet, and a spread of branches of 352 feet diameter.

There are many fine tall beeches in the park of Earl Bathurst at Cirencester, of which Plate 1 gives a good idea, and shows the reproduction from seed in this part of the park to be very good, though a considerable number of other trees, such as ash and sycamore, are growing as well or better than the young beeches under the shade of the tall ones, which in this view are not so remarkable for their size as for their clean cylindrical trunks.

At Ashridge Park, Bucks, the property of Earl Brownlow, are perhaps the most beautiful and best grown beeches in all England, not in small numbers, but in thousands. Though the soil is neither deep nor rich, being a sort of flinty clay overlying lime-stone, it evidently suits the beech to perfection, and in some parts of the park there is hardly a tree which is not straight, clean, and branchless for 40 to 60 feet, whilst in other parts, where the soil is heavier and wetter, and where oaks grow among the bracken to a great size, the beeches are of a more branching and less erect type.

The largest and finest beech, from a timber point of view, at Ashridge, known as the King Beech, was blown down about 1891, and was purchased for £36 by Messrs. East of Berkhampstead. Loudon says that this tree in 1844 was 114 feet high, with a clear trunk of 75 feet, which was 5 feet 6 inches in girth at that height. Evidently this was less than its real height. Mr. Josiah East tells me that as it stood it had about 90 feet of clean trunk, of which the lower 15 feet was partly rotten and not measured. The sound part was cut into three lengths as follows :—

17 feet × 29 inches, ¼ girth	.	.	.	= 99 cubic feet.
28 ,, × 25 ,, ,,	.	.	.	= 136 ,, ,,
30 ,, × 23 ,, ,,	.	.	.	= 110 ,, ,,
butt, say, 15 ,, × 36 ,, ,,	.	.	.	= 135 ,, ,,
90				480

The branches were partly rotten and much broken in falling, so that they were only fit for firewood. But the celebrated Queen Beech remains, and though in one or two places it shows slight signs of decay, it may, I hope, live for a century or more, as it is in a fairly sheltered place, and has no large spreading limbs to be torn off by the wind. This extremely perfect and beautiful tree was photographed with great care from three positions by Mr. Wallis (Plate 3), and as carefully measured by Sir Hugh Beevor and myself in Sept. 1903. We made it as nearly as possible to be 135 feet high (certainly over 130), and this is the greatest height I know any deciduous tree, except the elm, to have attained in Great Britain. Its girth was 12 feet 3 inches, and its bole straight and branchless for about 80 feet, so that its contents must be about 400 feet to the first limb.[1] Other extraordinary beeches at Ashridge are figured. Plate 4 is an illustration of natural inarching of a very peculiar type: the larger tree is 17 feet 6 inches in girth, the smaller, 4 feet 9 inches, and the connecting branch 12 feet long. It passes into the other tree without any signs to indicate how the inarching took place, and might almost have been a root carried up by the younger tree from the ground, as it has no buds or twigs on it. There are several beeches at Ashridge with very large and curious bosses on the trunk; one of these (Plate 5) at the base measured 21 feet over the boss, another had a large burr growing out of the side of a straight, clean, healthy tree at 40 feet from the ground. Such burrs are formed on the trunks of healthy as well as of diseased beeches, but I am not sure whether they ever have their origin in injuries produced by insects, birds, or other extraneous causes. Sometimes they have a horny or almost coral-like growth. Such burrs when cut through have an ornamental grain, which might be used for veneers when sufficiently compact and solid, but are left to rot on the ground by timber merchants, who as a rule place no value on such products.

In some parts of this park the beeches show a remarkably wide-spreading network of snake-like roots on the surface, which, though not uncommon in this tree when growing on shallow soil, are here unusually well developed. There is a remarkable beech clump to the east of the house containing 26 trees in a circle of 197 paces (11 of them grow in a circle of 78 paces), of which every tree is large, clean, and straight. The largest of them is about 125, perhaps 130, feet high, and 13 feet 10 inches in girth, and the average contents of the trees probably over 200 feet. I do not think I have ever seen in England such a large quantity of timber on so small an area.

But though it is doubtful whether any place in England can boast so many perfect beech trees as Ashridge, this park contains also some of the finest limes, the largest horse-chestnuts, and the most thriving and bulky chestnuts; and in a wood not far off is an ash which is much the best-grown tree of its species, if not the largest, that I have seen in England. All things considered, I doubt whether there is a more interesting and beautiful type of an English park than Ashridge, for though it contains few exotic trees, and no conifers except some Scotch pines, it has a magnificent herd of red, of Japanese, and of fallow deer, as well as flocks of St. Kilda sheep and of white Angora goats.

At Rotherfield Park, Hants, there is an immense pollard beech, of which I have a photograph kindly sent me by the owner, Mr. A. E. Scott, who gives its girth as 28 feet 3 inches at the narrowest point, 3 feet from the ground.

---

[1] According to Loudon, iii. 1977, this tree was in 1844 110 feet high, 10 feet in girth at 2 feet, and 74 feet to the first branch.

At Slindon Wood, near Petworth, Sussex, between the South Downs and the sea, which is seven miles distant, on the property of Major Leslie, there was in 1903 one of the finest beech woods in England, growing on chalk soil, of which I have particulars from Mr. C. H. Greenwood, and of which I give an illustration from a photograph sent me by him (Plate 6). Mr. Greenwood states that 634 trees were recently cut and sold in this wood, many of them being 70 and several 80 to 90 feet long to the first limb, and quarter girthing 20 inches in the middle. One tree now standing measures, without the top, 70′ × 26″ = 320 feet, and on one acre at the east side of the wood are standing 60 which would average 150 feet each, making 9000 cubic feet to the acre. The tallest tree is 90 feet to the first bough, with 21 inches ¼ girth = 275 feet. This is perhaps the largest yield of beech per acre of which I have any record in England.

In Windsor Park there are some fine old beeches, of which three are figured by Menzies.[1] His plate 4 shows a remarkable old pollard at Ascot Gate 30 feet in girth, which he supposed to be 800 years old, and another, his plate 6, on Smith's lawn, of similar age and 31 feet 9 inches in girth. The third, Queen Adelaide's Beech, is a tree of no great size or beauty. It measured in 1864 8 feet 6 inches in girth, when supposed by Menzies to be 140 years old. In 1904 it had only increased 10 inches in girth. The finest beech now growing at Windsor —Mr. Simmonds, the deputy-surveyor of the Park, who was good enough to show it to me, agrees in this—is a tree near Cranbourne Tower, which in March 1904 measured 125 feet by 15, with a fine clean bole, but not equal to that of the Queen Beech at Ashridge.

The two largest beech trees, of whose measurement I have exact particulars, were both blown down in the heavy gale of September 1903, I believe on the same night. One of these was at Cowdray Park in Sussex, the property of the Earl of Egmont, and grew on sandy soil near the top of the great chestnut avenue at a considerable elevation, perhaps 400 feet. I saw it lying on the ground not long after, and obtained from Mr. Barber, steward on the estate, the following careful measurements :—

Butt 22 feet by 72 inches ¼ girth = 792 feet. Limbs measured down to 9 inches ¼ girth only, 43 in number, contained 924 feet 6 inches. Total 1716 feet 6 inches. Measured on the ground 21st September 1903.

The other was the great beech at Cornbury Park, of which I give a photograph taken after its fall (Plate 7), that gives an idea of its immense size. I saw the stump of this tree two years afterwards, and counted about 230 rings in it, which justify the belief that it may have been planted by Evelyn. Mr. C. A. Fellowes, agent for the property, had the tree carefully measured after its fall, and gives its height as 120 feet, girth 21 feet 4 inches. Cubic contents 1796 feet (nothing under 6 inches quarter girth being measured).

A magnificent beech growing in Studley Park, the seat of the Marquis of Ripon, was figured by Loudon, iii. 1955, and is there stated to have been 114 feet high.

[1] *History of Windsor Great Park and Windsor Forest*, 1864.

Mr. O. H. Wade, agent for the estate, tells me that this tree cannot now be identified.

Another celebrated tree, mentioned by Loudon as Pontey's Beech, was measured for him in 1837 by the direction of the Duke of Bedford in the Park at Woburn Abbey. It was then 100 feet high, with a clean bole of 50 feet, and was 12 feet 6 inches in girth at 4 feet. When visited in July 1903 it was about the same height and 14 feet 6 inches in girth, and was estimated to contain nearly 600 cubic feet.

A tree known as the Corton Beech at Boyton, Wilts, once the home of Mr. Lambert, author of the *Genus Pinus*, and mentioned by Loudon as one of the largest in England, was blown down a few years ago, and I have not been able to get its dimensions.

There were some very fine beeches at Castle Howard, Yorkshire, the seat of the Earl of Carlisle, one of which Loudon gives as 110 feet by 14 feet 2 inches, with a clean bole of 70 feet, and the other as containing 940 feet of timber, but when I visited this fine place in 1905 I could not identify either of these trees as still standing, though I saw many in Raywood of great size, with clean boles of 50 to 60 feet. A tree standing outside the garden wall was remarkable for the very rugged bark on its trunk, which up to 8 to 10 feet from the ground was more like that of an elm than a beech.

In Scotland, though the beech does not attain quite the same height and size as in some parts of England, it is a fine and commonly planted tree.

The self-layered beech at Newbattle Abbey near Dalkeith, the property of the Marquess of Lothian, eight miles from Edinburgh, must be looked on as the most remarkable, if not the largest, of all the beeches of the park or spreading type now standing in Britain; and though difficult to represent such a tree by photography in a manner to show its great size, every pains has been taken by Mr. Wallace of Dalkeith to do it justice (Plates 8 and 9). This splendid tree is growing in light alluvial soil in front of the house, and not far from the banks of the North Esk river, and may be 300 years old or more. It was in Loudon's time 88 feet high, and the trunk 9 feet in diameter (probably at the base), with a spread of branches of 100 feet. When I visited it in February 1904 under the guidance of Mr. Ramsay, who has known the tree for many years, I made it about 105 feet high, with a girth at about 5 feet—which is near the narrowest part of the bole—of 21 feet 6 inches. The trunk, as will be seen from the figure, is unusual in shape, and shows no sign of decay except where one large limb has been blown off, and this has been carefully covered with lead. But the numerous branches which have drooped to the ground, taken root, and formed a circle of subsidiary stems round the main trunk, are its most peculiar feature, and may remain as large trees for centuries after the central stem decays. The first of these has produced 7 stems of various sizes growing into fresh trees, at a distance of 8 to 12 yards from the trunk. The second has 2 large and 3 smaller stems. The third has 3 large stems about 30 to 40 feet high and 3 to 4 feet in girth. The fourth has 3 large and 6 smaller ones. The fifth is not yet firmly rooted, but is fastened down in several places to prevent the wind from moving it. The total circumference of these branches is about 400 feet.   Detailed measurements by Mr. Ramsay are given below.[1]

A similar instance of self-layering, perfectly natural, was to be seen in the Kew Gardens, where a very fine beech, though by no means such a giant as the Newbattle tree, was surrounded by a fence in order to protect it.   This tree, however, having become seriously decayed, had its main stem cut down in 1904.

Among the best specimens I have seen in Scotland are those at Hopetoun House, near Edinburgh, the seat of the Marquess of Linlithgow, where I measured a tree 110 feet high, with a clean bole of about 50 feet, and a girth of 12 feet.   At Blair Drummond, near Perth, the seat of H. S. Home Drummond, Esq., Henry measured one of 117 feet high by 16 feet 6 inches in girth, and at Methven Castle, the seat of Colonel Smythe, another which is 120 feet high by 17 feet 2 inches in girth.   This tree divides into three stems at about 20 feet, and is the tallest of which we have any certain record in Scotland.   At Gordon Castle is a very fine beech with spreading roots (Plate 10) measuring 95 feet by 15 feet 8 inches.   At Castle Menzies, Perthshire, the property of Sir Neil Menzies, is a very fine beech, which is described by Hunter[2] as a vegetable "Siamese Twins."   Whether originally two trees or one is difficult to say, but it seemed to me to be from a single root which had forked a little above the ground and then grown together again, leaving an opening through which Hunter says an ordinary sized person might pass, but which in 1904 was smaller.   At Inverary Castle is another example of an inosculated beech, known as the Marriage Tree, which, from a photograph published by Valentine, does not seem to be so striking as the one at Castle Menzies.

[1] Newbattle Abbey, Midlothian, N.B.   Measurement of the great beech tree, August 25, 1903, by Mr. John Ramsay. Girth in feet, inches, etc., of trunk—

At the ground	.	.	.	.	43 feet 8 inches.
About 1 foot up	.	.	.	.	37 ,,
,, 2½ feet ,,	.	.	.	.	27 ,, 8 ,,
,, 3 ,, ,,	.	.	.	.	25 ,, 9½ ,,
,, 4 ,, ,,	.	.	.	.	23 ,, 1½ ,,
,, 4½ ,, ,,	.	.	.	.	21 ,, 11½ ,,
,, 5 ,, ,,	.	.	.	.	20 ,, 3½ ,,
,, 6 ,, ,,	.	.	.	.	19 ,, 7½ ,,

The ground measurement was taken by allowing the tape to lie on the roots as near to the uprising of the buttresses as possible, and is necessarily vague.   The measurement at 6 feet up is the most correct, being taken on a line marked at intervals all round with white paint for future comparison.

Circumference of foliage fully 400 feet; diameter of foliage averages 130 to 140 feet; height, 112 feet.

The following are a few of the branches with the girth of them, and the girth of the branches springing up from the main branches rooted in the ground :—

No 1.—Branch girth, 1 foot 10 inches, with two branches growing up from it; girth of both these new branches, 4 feet 5 inches each.

No. 2.—Branch girth, 1 foot 8 inches, having three branches springing up from it, one 5 feet 5 inches, one 5 feet 1 inch, one 23 inches by 1 foot 11 inches in girth.

No. 3.—Branch girth, 12½ inches, having three branches springing up from it, one 4 feet 7½ inches, one 24½ inches, one 4 feet 4 inches in girth.

No. 4.—Branch girth, 12 inches, with two branches springing up from it, one 2 feet 8½ inches, one 12 inches in girth.

No. 5.—Branch girth, 1 foot 7 inches, with three branches springing up from it, one 2 feet 4½ inches, one 12 inches, one 18 inches in girth.

No. 6.—Branch girth, 2 feet 4 inches, with five branches springing up from it, one 4 feet 4 inches, one 3 feet 8 inches, one 4 feet, one 3 feet 4 inches, one 1 foot 11 inches in girth.

[2] Hunter, *Woods, Forests, and Estates of Perthshire*, 1883, p. 397.

There are two beeches standing on a mound near the road to Lochfynehead in the Park at Inverary, which are known as the Doom trees, because in former times they were said to have been used as a gibbet for criminals; the largest of them measures 75 feet by 16 feet 5 inches. The Duke of Argyll, however, doubts this tradition.

There is another very fine beech, the largest I know of in the West Highlands, at Ardkinglas, at the head of Lochfyne, under which Prince Charles's men are said to have camped in 1745. Though of no great height it has a girth of 18 feet 8 inches, and spread of branches 30 yards in diameter.

In Ayrshire the largest beech is at Stair House. According to Renwick,[1] in 1903 it was 100 feet high, and 18 feet 9 inches at 4 feet 3 inches above the ground. At Kilkerran, in the same county, Renwick records a beech 21 feet 3½ inches at 3 feet from the ground, which, however, had a bole of only 4 feet. Other large beeches in Scotland occur at Eccles in Dumfriesshire and at Belton in East Lothian. The Eccles Beech, according to Sir R. Christison, was little inferior to the Newbattle Beech; according to Hutchinson, in 1869 it was 20 feet in girth at 4 feet up. I learn from Dr. Sharp that it has been dead for some years. The Belton Beech in 1880 was 20 feet 4 inches girth at 5 feet, with a 13-feet bole and a height of 63 feet.

One of the most striking effects produced by the beech in Scotland is the celebrated beech hedge of Meikleour, in Perthshire, on the Marquess of Lansdowne's property. An account of this hedge is given in the *Gardeners' Chronicle*, Dec. 15, 1900. This hedge forms the boundary between the grounds and the highway, and has to be cut in periodically, which is done by men working on a long ladder, from which they are able to reach with shears to about 60 feet. Local history says that this hedge was planted in 1745, and that the men who were planting it left their work to fight at the battle of Culloden, hiding their tools under the hedge, and never returning to claim them.[2] It is 580 yards long, and composed of tall, straight stems planted about 18 inches apart, and nearly touching at the base. The average height of the trees, as I am informed by Mr. Donald Matheson, is 95 feet, and their average girth at 3 feet is 18 to 36 inches. He adds that "close to the ground they are as fresh and green as a young hedge." An illustration of this hedge, taken specially for our work by Mr. D. Milne of Blairgowrie, gives a good idea of its appearance in October 1903 (Plate 11).

I am informed by Sir Herbert Maxwell, M.P., that a remarkably similar occurrence is on record at Achnacarry, on the property of Cameron of Lochiel; here the trees were laid in ready to plant in 1715, and the men were also called off to take part in the rebellion of that year. The trees were never planted, and have grown up in a slanting position close together just as they were left.

In a paper on the "Old and Remarkable Trees of Scotland," published in 1867 by the Highland and Agricultural Society of Scotland, many other remark-

[1] Renwick in *British Association Handbook*, p. 140 (1901). We are much indebted to Mr. John Renwick for measurements and descriptions of large and interesting trees in the south-west of Scotland.

[2] Hunter, *loc. cit.* 379.

able beeches are mentioned, of which one at Edenbarnet in the parish of Old
Kilpatrick, Dumbartonshire, is said to be 140 feet high ; but the measurements of
many of the trees in this compilation are so unreliable that I cannot believe them
without confirmation.

J. Kay, in *Scottish Arb. Soc. Transactions*, ix. p. 75, mentions a tree in the Beech
Walk at Mount Stuart in Bute, which in 1881 was 120 feet by 11 feet 9 inches, with a
clean bole 60 feet high, and contained 450 feet of timber.

In Ireland the beech is probably not a native tree. According to Hayes[1] it
was first introduced at Shelton, near Arklow, where, in 1794, there were beech
trees as much as 15 feet in girth, and many carrying à girth of 10 feet for more than
40 feet high. Another growing at Tiny Park was 16 feet 3 inches in girth, and
continued nearly of that girth for 36 feet. Hayes also mentions, as an instance of
the rapid growth of the beech in Ireland, "several at Avondale, which were transplanted within thirty years on a swelling ground at that time much exposed to
storm, are now (1793) from 7 feet 6 inches to 6 feet 6 inches at a foot from the
ground, and continue nearly of that size from 8 to 20 feet in height. Of two which
were planted in a richer soil near the river, and are now (1793) just fifty-four years
from the mast, one measures 9 feet round, the other 9 feet 6 inches."

The finest beeches in Ireland, probably, are those occurring at Woodstock (Co.
Kilkenny), the seat of E. K. B. Tighe, Esq.—a property which is remarkable all
round for magnificent trees of many kinds, and which is in the possession of a family
that for generations has been deeply interested in forestry and arboriculture. The
measurements of many trees have been taken periodically for nearly a century.
The best beeches on this beautiful property occur in the meadow land by the River
Nore, close to the village of Inistioge. The following table gives an interesting
series of measurements of these beeches :—

No.	GIRTH.						HEIGHT.	
	1825.	1830.	1834.	1846.	1901.	1904.	1901.	1904.
	ft. in.	ft. in.	ft. in.	ft. in.	ft. in.	ft. in.		
A_3	10 9	11 1	11 6	12 6	20 6	20 7	81	86
C_7	...	...	...	12 7	17 3	17 9	97	99
B_3	...	...	...	12 5	14 0	...	91	...
B_5	12 1	12 10	14 4	15 4	18 9	18 10	113	117
B_6	11 10	12 3	13 8	14 10	17 9	...	108	...
B_2	11 4	11 11	12 10	13 8	15 8	16 4	112	109
B_1	11 0	11 6	12 7	13 8	16 6	...	106	...
B_9	11 9	...	12 9	14 0	16 7	...	120	...
B_8	...	...	9 5	10 1	12 3	...	100	...

The measurements up to 1901 are from the foresters' records ; those of 1904
were taken by Henry. The beech A_3 has a great bole, dividing into three limbs
at 18 feet up, and is a very wide-spreading tree. C_7 is pressed on each side by two
lime trees, and is narrow in shape. The most remarkable of all is B_5, which is
probably the tallest beech in Ireland.

[1] Hayes, *A Practical Treatise on Planting* (1794), pp. 109, 118.

As showing the rate of growth of the beech in Co. Galway, a beech measured by Lord Clonbrock at Clonbrock was 11 feet 3 inches in girth in 1871, and 15 feet in 1903. A beech hedge at Kilruddery, Co. Wicklow, the seat of Lord Meath, said to be 300 years old, was measured by Henry in 1904, when it was 18 feet through and 29 feet high. It is clipped regularly, and forms a dense, impenetrable mass.

BEECH COCCUS

We are indebted to Mr. R. Newstead of the Grosvenor Museum, Chester, for particulars of the coccus which in some seasons, and in certain parts of England, has been of late years very injurious to the beech. A fuller account of this insect has been written by him in *Journ. R. Hort. Soc.* 1900, vol. xxiii. p. 249, and in a leaflet recently published by the Board of Agriculture. From this we take the following precis :—

The trunks and, less frequently, the main branches of good-sized beech trees are often covered, to a greater or less extent, with irregular spots of a white cottony substance. The latter is really the covering of white felted wax fibres secreted by the felted beech coccus (*Cryptococcus fagi*, Bärensprung), a minute, hemispherical, lemon-yellow insect, about one twenty-fifth of an inch long, without legs, but furnished on the underside with a well-developed beak, which it buries in the bark for the purpose of sucking up the juices of the tree. When once a tree is attacked the number of individuals of the pest becomes in time so great that it is doubtful whether a badly-infected tree ever recovers unless active measures be taken against the insect. The waxy covering of the latter is sufficient to protect it against the effects of any of the insecticides usually applied by spraying, and its habit of preferring the deepest part of the fissures in the bark makes it difficult to remove with certainty. The only remedy at all likely to succeed is that of thoroughly scrubbing the bark with a stiff brush and soap and water, the latter mixed in the proportion of half a pound of soft soap to each gallon of water ; and the success of this treatment depends for the most part on the amount of care taken to dislodge the insects by means of the brush.

TIMBER

The timber of the beech is not valued so highly in England as abroad, where it is considered as the best fuel in general use, and is little used in carpentry or building, as it is hard, brittle, and liable to be attacked by beetles. It weighs when green about 65 lbs. to the cube foot, when dry about 50. Its durability is said to be increased by seasoning it in water, and it is more durable when entirely under water than most timbers, being highly recommended by Matthews and Laslett for planking the sides and bottoms of ships. In France it is used, when creosoted, for railway sleepers, but requires more than twice as much creosote to preserve it as oak does, and is not used in England, so far as I know, for this purpose. It is also used for tool handles, rollers, butchers' blocks, brush heads, planes, and general turnery, but decays rapidly when exposed to the weather.

The principal centres for beechwood furniture in England are at High Wycombe, and Newport Pagnell in Bucks, and the price of clean trunks in these districts is from 1s. to 1s. 6d. per cube foot standing, according to the situation. Beechwood is also used largely for making saddle-trees, and in consequence of the great demand for these during the South African war, went up to a very high price in 1901, when I was offered 1s. 4d. a foot standing for beech trees which in ordinary times would not be worth more than 8d. or 9d. a foot. Being easy to split it is, where there is a demand for firewood, easier to dispose of the branches and rough parts of the tree for this purpose, but the amount of waste is much greater in the beech than in some other trees, unless grown in thick woods. For more minute particulars of the characters and uses of this timber, Stone's *Timbers of Commerce*, p. 231, and Loudon, pp. 1959-64, may be consulted with advantage.   (H. J. E.)

AILANTHUS

Ailanthus, Desfontaines, *Mém. Acad. Paris*, 1786 (1789), 263, t. 8 ; Bentham et Hooker, *Gen. Pl.* i. 309 (1862) ; Prain, *Indian Forester*, xxviii. 131, Plates i. ii. iii. (1902).

LOFTY trees with very large alternate imparipinnate leaves. Flowers small, polygamous, bracteolate, in panicles. Calyx 5-toothed, imbricate. Petals 5, valvate, disk 10-lobed. Stamens 10 in the staminate flowers, 2-3 in the hermaphrodite flowers, and absent in the pistillate flowers. Ovary present in pistillate and hermaphrodite flowers, rudimentary in staminate flowers, deeply 2-5 cleft with connate styles: ovules 1 in each cell. Fruit of 1-5 samaras, with large membranous wings, each samara containing 1 seed.

Ailanthus belongs to the Natural order Simarubeæ, and consists of about eleven species occurring in India, Indo-China, China, Java, Moluccas, and Queensland. Most of the species are tropical trees, *Ailanthus glandulosa* being until lately the only species which was known to occur in temperate regions ; but *Ailanthus Vilmoriana*, Dode,[1] must be here mentioned. This is a tree remarkable for its prickly branchlets, of which only one specimen is known, namely, a young, healthy, vigorous tree grown in M. de Vilmorin's garden at Les Barres.[2] It was raised from seed sent by Père Farges in 1897 from the mountains of Szechuan in Central China ;[3] and is certainly a very distinct species. I saw it in the summer of 1904, and in general aspect there is little to distinguish it from the common species. It is now about 20 feet in height. The leaflets in this species are less abruptly acuminate, not falcate, much duller above and paler beneath, with larger glands than in *Ailanthus glandulosa*. All the parts of the tree are much more pubescent than in that species.

Ailanthus grandis,[4] Prain, a new species from Sikkim and Assam, which attains 120 feet high, may be here mentioned, as it is possible that it might be grown in Cornwall or in Kerry. It has not yet been introduced.

---

[1] *Revue Horticole*, 1904, p. 445, fig. 184.

[2] Figured in *Fruticetum Vilmorinianum*, 1904, p. 31 ; where it is called *Ailanthus glandulosa*, var. *spinosa*.

[3] Mr. E. H. Wilson informs us that it is very common in the valleys of the Min, Tung, and Fou rivers, between 2000 and 4500 feet. He says that it is much more spiny in the young than in the adult state, and that it has much larger foliage than the common species. A plant is now growing at Kew, and is referred to by Mr. Bean in *Gardeners' Chronicle*, xxxviii. 276 (1905).

[4] *Indian Forester*, xxviii. 131, Plate i. (1902).

AILANTHUS GLANDULOSA, AILANTHUS TREE

Ailanthus glandulosa, Desfontaines, *Mém. Acad. Paris.* 1786 (1789), 263, t. 8 ; Loudon, *Arb. et Frut. Brit.* i. 490 (1838) ; Britton and Brown, *Illustrated Flora of the Northern United States and Canada*, ii. 355, Fig. 2272 (1897).

A tree attaining 100 feet in height and 13 feet in girth ; branches massive and forming an oval crown, which becomes flattened at the top in old trees. Bark smooth, grey, or dark brown, and marked by longitudinal, narrow, pale-coloured fissures, which are very characteristic.

Leaves deciduous, compound, 1-3 feet long, imparipinnate, with 7-9 (sometimes even 20) pairs of leaflets, which are either opposite or nearly so, shining above, pale and glabrous (occasionally slightly pubescent) beneath, and unequally divided by the midrib. Each leaflet is stalked, ovate, or ovate-lanceolate, acute or acuminate at the apex, cordate or truncate at the base, entire in margin, except that near the base there are 1-4 pairs of glandular teeth. Stipules absent. The leaves appear late in spring, and exhale when rubbed a disagreeable odour which renders them distasteful to animals. They fall off late in autumn, absciss layers being formed at the base of the leaflets as well as of the main stalk ; the former usually drop first.

Flowers appearing in July and August in large panicles at the summit of the branchlets, either unisexual or hermaphrodite ; but as a rule the trees are practically diœcious, and those bearing staminate flowers give off an objectionable odour.

Fruit, 1-5 keys, resembling those of the ash, linear or oblong, membranous veined, with a small indentation above the middle on one side, close to where the seed is located ; and the wings on both sides of the seed are slightly twisted, so that the fruit in sailing through the air moves like a screw. The keys are bright red or purplish brown in colour, and are very conspicuous amidst the green foliage.

Seedling : the cotyledons appear above the soil on a caulicle about an inch long and are foliaceous, coriaceous in texture, oboval, obtuse, shortly stalked, entire in margin, and pinnate in venation. The stem above them is pubescent, and at a short distance (about $\frac{1}{2}$ inch) up bears two leaves, which are trifoliolate and long-stalked, the terminal leaflet being lanceolate, acuminate, and entire, the two lateral shorter and toothed.[1] Higher up ordinary pinnate leaves are borne. Plate 15 A shows a seedling raised by Elwes from seed ripened on a tree overhanging Dr. Charles Hooker's garden at Cirencester in 1900 ;[2] sown November 26, germinated under glass in May 1901, and photographed on August 28 of the same year, when it measured about a foot high ; the roots, which were very succulent and brittle, were 13 inches long. The seedlings were planted out in May 1902, and grew very rapidly, attaining 5 feet in height, but did not ripen their wood, which was killed back in some cases nearly to the ground. They are now (January 1905) 4-6 feet high.

[1] See Plate 14, fig. B.

[2] As I know of no other tree in the neighbourhood this case seems to confirm Bunbury's observation that the tree in some cases is capable of self-fertilisation.—(H. J. E.)

IDENTIFICATION

In summer the Ailanthus is readily distinguished from all other trees cultivated in England by its large pinnate leaves, which have *at the base of the leaflets on each side one or two glandular teeth*. The black walnut, butternut, and *Cedrela sinensis* have somewhat similar foliage; but in these the glandular teeth are wanting. The bark of Ailanthus is quite peculiar, and when once seen cannot be confounded with that of any other tree.

In winter Ailanthus is easily recognised by its bark in trees of a certain size; but in all stages of growth it is well marked by the characters of the buds and branchlets.

The buds are alternate, uniform in size, small and hemispherical, and show externally 2 or 3 brown tomentose scales.[1] The buds are set obliquely on the twigs just above the leaf-scars. The latter are large, heart-shaped, and slightly concave; and on their surface may be seen about 7 little elevated cicatrices which correspond to the vascular bundles of the fallen leaves. No true terminal bud is formed; and at the apex of the twig there is an elevated small circular scar, which marks the spot where the tip of the branchlet fell off in summer. The twigs are very coarse, glabrous, or finely pubescent, shining and brown in colour, with a few plainly visible lenticels. The pith is large, buff or yellowish in colour, showing clearly on section the medullary rays. In Cedrela there is a large terminal bud, and the leaf-scar has 5 cicatrices. The chambered pith of Juglans will readily distinguish the black walnut and butternut.

VARIETIES

Several varieties are mentioned in books; *aucubæfolia, pendulifolia, rubra*, and *flavescens* being recognised by Schelle;[2] but it is doubtful if any of these are sufficiently marked to deserve recognition. The *Ailanthus flavescens*[3] of gardens was determined by Carrière to be *Cedrela sinensis*. A form with variegated leaves is mentioned by Koch,[4] but it is exceedingly rare. The Kew Hand-list only admits one variety, *pendula*, a form somewhat weeping in habit.

DISTRIBUTION

Ailanthus glandulosa has been only found truly wild on the mountains of the province of Chihli in Northern China; but it is cultivated in most parts of China, and doubtless was once a constituent of the forests of the northern coast provinces, most of which have been destroyed by the Chinese. I never saw it wild in any of the mountain forests of Central or Southern China. When first introduced

---

[1] A plate showing buds will appear in a later part. [2] *Laubholz-Benennung*, 279 (1903).
[3] See article on the "Ailanto or Tree of Heaven" by Nicholson, in *Garden*, 1883, xxiv. 63, with figure of flowers, fruit, and foliage, and many interesting details concerning propagation, etc.
[4] Koch, *Dendrologie*, i. 569 (1869).

into Europe it was supposed to be the species of Rhus which yields Japanese varnish or lacquer; and even now it is often called in France *Vernis du Japan*. The tree, however, is unknown wild in Japan, and is seldom or never cultivated there. The Chinese in classical times were well acquainted with Ailanthus, which they called *ch'u*, a word explained as meaning "useless wood," as it was in ancient times (as well as at present) used only for firewood.[1] Popularly Ailanthus and Cedrela are now called *ch'un* trees, the former being distinguished as the "stinking ch'un," and the latter as the "fragrant ch'un."

In China the Ailanthus grows to be a large tree; but the timber is little valued. The root-bark is used, as a strong infusion, in cases of dysentery.[2] In the Pharmaceutical Museum, London, there are several specimens of barks bearing the Chinese name for Ailanthus; but these are doubtfully referable to that species; and the whole subject of the use of Ailanthus bark for dysentery requires further investigation.[3]

In the Kew Museum there are specimens of silkworms (*Attacus Cynthia*, Drury), which feed on the leaves of Ailanthus in North China; and there are also samples of the "wild silk" produced, which is made into one kind of pongee. This species of silkworm was introduced into France in 1858; and large numbers of Ailanthus trees were planted with a view to the feeding of the silkworms. The winter of 1879 killed off all the silkworms; and apparently the cultivation of the tree in France for the production of silk is a thing of the past.

In the Kew Museum there is a note attached to a specimen of the wood of *Ailanthus glandulosa* from Tuscany, which says that the bark yields a resinous juice; but there is no account of such a resin from Chinese sources; and exudation from the bark has not been observed in trees growing in England or in France. In India, however, the resin, called *muttee-pal*, is derived from the bark of *Ailanthus malabarica*, and is used both as an incense and as a remedy for dysentery.

Introduction

Ailanthus glandulosa was first introduced from China in 1751. In *Hortus Collinsonianus*,[4] p. 2, a memorandum is copied which was left by Collinson, stating : "A stately tree raised from seed from Nankin in 1751, sent over by Father d'Incarville, my correspondent in China, to whom I sent many seeds in return; he sent it to me and the Royal Society." Père d'Incarville[5] was a French Jesuit missionary, who died at Peking in 1757. In *Trans. Phil. Soc.*, 1757, a paper is printed, which was read on 25th November 1756, being a letter from John Ellis to P. C. Webb; and it mentions two trees which were growing, one in Webb's

[1] In the *Shu-Ching*, it is said : "In the ninth month they make firewood of the *ch'u* tree."

[2] On the therapeutical value of this drug, see articles by Drs. Dudgeon and Robert, in *London Pharmaceutical Journal*, ser. iii. iv. 890, and vii. 372.

[3] The bark has been found to be an excellent vermifuge in cases of tapeworm. See Hetet, in *U.S. Dispens.* 15th edition, 1564.

[4] Compiled by L. W. Dillwyn, and published at Swansea in 1843.

[5] In Cibot, *Mém. Conc. Chinois*, ii. 1777, 583, d'Incarville's "Mémoire sur les vers à soie sauvage" is published, in which he speaks of the Ailanthus as the *frêne puant* (stinking ash) of North China.

garden at Busbridge, near Godalming, and another in the Chelsea Physic Garden, both raised from the seed sent by Père d'Incarville. The tree is here first described as *Rhus sinense foliis alatis, foliolis oblongis acuminatis ad basin subrotundis et dentatis.*[1]

TREE OF HEAVEN

This name is often given to the tree in England, corresponding to the German *Götterbaum.* It is not the translation of any Chinese name, as has often been erroneously stated. Desfontaines' original description occurred in a rare book which has not been looked up by most writers on the tree. He was well aware that the tree came from China, but in selecting a name for the genus he took it from another species which he found figured in Rumphius' *Hortus Amboinensis,* v. cap. 57, tab. 132. This species, left undescribed by Desfontaines, is *Ailanthus moluccana.* Rumphius calls it *arbor cœli,* the equivalent of the native name in the Amboyna language, *Aylanto,* which signifies "a tree so tall as to touch the sky." "Tree of Heaven" is accordingly a translation of the name of Rumphius, and is more properly applied to the tall tropical species than to *Ailanthus glandulosa,* which does not attain any remarkable height.

CULTIVATION

The Ailanthus is easily propagated from seeds; but as trees bearing male flowers are objectionable on account of their odour, it is preferable to propagate the tree from root-cuttings obtained from female trees. In addition to the disagreeable odour of the male flowers, there may be some foundation for the belief prevalent in the United States that they cause stomachic disturbance and sore throat. The pollen from staminate flowers, doubtless, occasions a kind of hay fever.

The tree suckers freely from the root and to a great distance, as far as 100 feet from the parent stem. At Kew these suckers frequently appear between the tiles of the floor of one of the buildings near which an Ailanthus stands. At Oxford[2] a root-sucker sent up a flowering shoot, and, what is more remarkable, produced simple leaves, giving some support to the idea that plants with compound foliage originated from those with simple leaves. The tree has extraordinary vitality. Dr. Masters[2] gives an account of a tree which was cut down, the stump being left in the ground below the surface. Several years elapsed during which nothing was observed, but after about ten years suckers were seen coming up in a gravel path adjacent, and these, being traced, were found to issue from the old stump.

Ailanthus reproduces itself freely from stools, and the coppice shoots thus obtained are very vigorous.

It was long supposed that Ailanthus would succeed even on the worst soils, but this is an error. It only does well on permeable soils, which are fairly moist,

[1] In the herbarium of the British Museum there is a specimen labelled *Hort. Busbridge,* which is undoubtedly from the original tree. It was cut down in 1856 owing to the great amount of shade it produced near the house (*Gard. Chron.* 1857, p. 55). There is another specimen from Kew Gardens, 1779, showing that the tree was cultivated early there.
[2] *Gard. Chron.* 1887, ii. 364.

and for this reason it is successfully used to cover railway and road embankments in France. It will not grow well on compact clay or on chalky or absolutely poor soils. In England it has only been planted as an ornamental tree, and it is very suitable for planting in towns, as it is not injured by smoke and is free from insect attacks and fungous diseases. Though it suckers freely, this is no objection in streets, where the pavements or wheel traffic prevents them from making an appearance. The young shoots are often killed by frost, but this only serves to keep the tree within bounds without the use of the pruning knife. The Ailanthus only makes one shoot annually, late in the spring, which continues to grow till October or November, and this is the reason why it is spring tender, as the tips of the shoots do not become properly lignified. The tree, however, bears the greatest cold in winter, and was not injured by the severe frost of 1879.

The tree produces flowers in England when it is about 40 feet high; and it fruits pretty frequently, but the seeds are often infertile.

When the Ailanthus is cut back annually, it grows rapidly and produces foliage of enormous size, suitable for the so-called tropical garden. Leaves of plants so treated have measured as much as 4 feet long and 15 inches wide.

The Ailanthus succeeds in a great variety of climates, and is planted in regions so diverse as Northern India, the United States, France, Germany, and Italy. In France it has not been successful as a forest tree, as it is not a social species, and is speedily dominated by native trees, if it survives the seedling stage, when it is sensitive to spring frosts. In warmer climates it easily regenerates by seed, and in consequence has become naturalised in many parts of Europe (as on the arid slopes of Mount Vesuvius, where it stands very well the drought), and in the United States,[1] where it often runs wild in old fields. American writers praise the tree for the value of its wood and the rapidity of its growth, as it is said to make timber faster than any of the native trees that are used for firewood.

The wood is yellowish or yellowish green, and is not clearly distinguishable into well-marked heart and sap woods, though in old trees the centre of the stem becomes deeper in colour. The wood has a specific gravity of 0.6, and is easily worked, taking a good polish. It rives easily. It is used by wheelwrights as a substitute for elm and ash; but is inferior to these, as it does not possess their elasticity or their capability of resistance to fracture. It is said, however, to bear well alternations of dry and wet.

Mr. J. A. Weale of Liverpool, who has paid great attention to the study of timbers, and knows more about them than any one in the trade in this country, writes to us that this wood resembles that of the ash so closely in structure, that the only real difference between the two is in the large cellular compound pores which are formed in the Ailanthus, as shown in the microscopical section which he enclosed.

Elwes is assured by Prof. C. S. Sargent that it makes nice furniture, and he has a specimen from a large tree which was cut down in the Palace Gardens at Wells, Somerset, of which the timber was bought by Mr. Halliday, a cabinetmaker, for £8.

---

[1] Also in Southern Ontario. See Britton and Brown, *loc. cit.*

Remarkable Trees

The largest Ailanthus was that at Syon, which was 70 feet high in Loudon's time, and nearly 100 feet in 1880.[1] It is now dead.

At Kew a vigorous tree is growing in the garden behind the Palace, which measures 73 feet high and 8 feet in girth. Not far off a number of Ailanthus trees of varying size, but none very large, occurs in a group, and they seem to be root-suckers; probably one of the original trees was planted in this spot in the eighteenth century.

At Milton Rectory, Steventon, Berks, there are two trees of equal height (78 feet), one girthing 9 feet 1 inch, and the other 8 feet 6 inches. Both these trees bloom freely every year, producing fruit of a bright red colour on the south side of the trees; and the seeds, as they fall in the garden near hand, produce seedlings which are very vigorous.[2]

At the Mote, Maidstone, there are two large trees, one of which is 70 feet high and 8 feet in circumference.

At Linton Park, Maidstone, is a tree growing in a shrubbery which was nearly 80 feet high by 6 feet 6 inches in 1902.

At Broom House, Fulham, the residence of Miss Sulivan, is a tree 80 feet high, with a bole 9 feet long and 10 feet in girth, which divides into two main stems (Plate 13).

At Fakenham, Norfolk, Sir Hugh Beevor has measured a tree 75 feet by 8 feet 11 inches.

At Barton, Bury St. Edmunds, an Ailanthus which was planted in 1826[3] measured in 1904 55 feet high, with a girth of stem of 5 feet 2 inches. Bunbury says that it is perfectly hardy at Barton, and did not suffer in the least from the severe winter of 1860. It was $3\frac{1}{2}$ feet girth at 3 feet from the ground in 1862. It flowered abundantly in August of 1861, the greater part of the flowers being hermaphrodite, and a considerable number of fruits were formed, but all dropped off before coming to maturity. It fruited abundantly in 1868. Bunbury says, generally there is only one samara to each flower, but not unfrequently two or three; he never saw more than three.

At Belton Park, the seat of Earl Brownlow, is a fine specimen of the tree, for a photograph of which (Plate 14) we are indebted to Miss F. Woolward, who gives its height as 83 feet, and its girth as 6 feet. This seems to be the tallest tree recorded in England.

At Burwood House, Cobham, Surrey, the seat of Lady Ellesmere, Colonel H. Thynne has measured an Ailanthus 71 feet high by 10 feet 10 inches girth, which, though partly fallen down and supported by a prop, is still a fine tree.

The tree seems to require a climate which is at once both warmer and drier in summer than that of the northern and western counties of England, and we do not know of any trees of any great size now existing in Scotland, Ireland, or Wales, though Loudon states that there was one at Dunrobin Castle, Sutherlandshire, 43 feet high.

(A. H.)

[1] *Garden*, 1880, xviii. 629.
[2] The Rev. H. Hamilton Jackson kindly sent us this information in a letter dated Dec. 10, 1903.
[3] Bunbury, *Arboretum Notes*, 88.

SOPHORA

Sophora, Linnæus, *Gen. Pl.* 125 (1737); Bentham et Hooker, *Gen. Pl.* i. 555 (1865).

TREES, shrubs, or perennial herbs, with naked buds and imparipinnate leaves. Flowers papilionaceous, in simple racemes or terminal leafy panicles. Calyx five-toothed, imbricate. Stamens ten, not united together, or rarely sub-connate. Ovary short-stalked, with many ovules. Pod moniliform, indehiscent, or tardily dehiscent.

The name Sophora was taken by Linnæus from the Arabic word *Sophera*, which indicated some leguminous tree. The genus belongs to the tribe Sophoreæ (Natural order Leguminosæ, division Papilionaceæ) characterised by imparipinnate leaves and ten free stamens. There are about twenty-five species of Sophora, generally spread throughout the tropical and warm temperate regions of the globe. The only species of importance which attain to timber size are *Sophora japonica* and *Sophora platycarpa*. *Sophora macrocarpa* from Chile and *Sophora tetraptera* from New Zealand are shrubs or small trees, which are frequently cultivated in the southern counties of England, and do not come within the scope of our work, although they are said to attain a height of 50 feet in the wild state.

Sophora platycarpa, Maximowicz, in *Mel. Biol.* ix. 70 (1873), (*Fuji-ki* in Japan), only lately[1] introduced into cultivation in England; but in the United States, where it has been grown for some time, it is said to have proved hardier than *Sophora japonica*.[2] It is a tree of considerable size, occurring in woods in Japan on the side of Fusiyama and in Nambu. It is similar in leaves and flowers to *Sophora japonica*; and, as will be pointed out in our account of that species, has been probably confused with it by writers on Japanese trees. The leaves are larger than in *Sophora japonica*, the leaflets being 2 to 3½ inches long, alternate, acuminate, glabrous or nearly so. The flowers are ½ inch long, white, and loosely arranged. The main difference is in the pod, which is membranous, flat, narrowly winged on each side, and irregularly constricted.[3]

---

[1] There are two plants at Kew which were raised from seeds obtained in 1896 from Späth of Berlin. See *Mittheil. der Deut. Dendr. Gesell.* 1896, p. 27.

[2] A. Rehder in Bailey's *Cyclopedia of American Horticulture*, p. 1684 (1902).

[3] *Sophora shikokiana*, Makino, in *Tokyo Botanical Magazine*, 1900, p. 56 (Yuko-noki in Japan), is described as a species closely allied to *S. platycarpa*, and as being widely distributed throughout the mountain districts of Japan. It is said to be a tree of considerable size.

SOPHORA JAPONICA, SOPHORA TREE

Sophora japonica, Linnæus, *Mantissa* i. 68 (1767); Loudon, *Arb. et Frut. Brit.* ii. 563 (1838); Shirasawa, *Iconographie des Essences Forestières du Japon*, i., Text, p. 86, Plate 50 (1900).

A large tree, with a straight cylindrical stem of considerable height in some cases, but more often in cultivated examples dividing at no great distance above the base; branches tortuous, with pendent tips; crown of foliage, large, broad, and rounded in shape. Bark brown or greyish and scaly, fissured longitudinally, but to no great depth; on young shoots and older branchlets, smooth and dark green.

Leaves deciduous, alternate, unequally pinnate, with nine to fifteen leaflets, which are sub-opposite, oval, pointed at the apex, often ending in a short bristle, dark green and opaque above, glaucous beneath. In the ordinary cultivated form they are apparently glabrous, but with a lens minute hairs may be detected on both surfaces. The petiolules are velvety; the main stalk is greenish, swollen at the base, and slightly pubescent. In certain wild specimens from China they are green and not glaucous beneath; and in Hupeh a well-marked variety occurs, in which the under surface of the leaflets, the petiole, and young shoots are densely white pubescent.

Flowers in large, loosely branched terminal panicles. They are somewhat variable in colour; in Central China white, at Canton a bright yellow, in cultivation in England pale yellow, sometimes tinged with purple. Calyx small, bell-shaped, five-toothed. Corolla, standard large, obtuse, round, recurved; wings oval-oblong; keel semi-orbicular, rounded, and of the same length as the wings. Pod long-stalked, 1 to 2 inches long, glabrous, fleshy, compressed, with a beak at the apex, and constricted between the seeds, which are one to five in each pod, dark brown in colour and kidney shaped.[1]

In England the tree produces flowers regularly, late in the season, in August, September, and October, but seldom if ever fruits.

IDENTIFICATION

Sophora japonica is readily distinguished in summer by the leaves, the characters of which have been already given, and by the branchlets, which are angled, very smooth and dark green, both in the young shoots and those of the second year. When the young shoots are cut they emit a strong peculiar odour. In winter the characters of the buds and branchlets must be noted. The buds are spirally arranged on the shoots; solitary or in pairs, one placed above the other; naked, *i.e.* not surrounded by any true scales, and dark violet densely pubescent. They are

[1] *Seedling.*—Seeds sown early in the year at Colesborne produced two or three young plants, which showed the following characters in July:—Caulicle an inch or more in length, terete, green, glabrous, ending in a long whitish tap-root with numerous lateral fibres. Cotyledons oblong-spathulate, ⅜ inch long, entire, rounded at the apex, tapering at the base, sub-sessile, coriaceous, dark green and minutely pubescent above, pale green below. Stem white appressed pubescent, giving off alternately about six compound leaves; the lower three with five leaflets, the terminal leaflet being larger and broader in proportion to its length than the others; the upper three with 7 to 9 leaflets, uniform in size and shape; all the leaflets oval, entire, shortly-stalked, their under surface with a scattered appressed pubescence, dense on the midrib. Small ovoid densely pubescent buds are produced, one in the axil of each leaf, the shoot being terminated by an oblong white pubescent larger bud.

very small and lodged in the leaf-scar, which is oval, with the bud in the centre, and displays three crescentic small cicatrices left by the vascular bundles of the petiole.   The leaf-scar is set obliquely on a projecting leaf-cushion.   The branchlets in winter are the same as in summer, but they show more clearly their zig-zag nature, and at their apex will generally be seen a little stub which indicates the point where the end of the branchlet fell off in summer, no true terminal bud being developed.   Occasionally a true terminal bud may be seen at the apex of the shoot, which is open and not concealed in the leaf-scar, minute, bearing two scales outwardly, and very pubescent.[1]

Varieties

In addition to the pubescent form of Central China, not yet introduced, a few varieties occur, concerning the origin of which little is known.

Var. *variegata.*—Leaves dull yellowish white in patches.   This form is neither robust in growth nor attractive in appearance.

Var. *pendula*[2] (Weeping Sophora). — One of the most formal of weeping trees.   It is usually grafted by budding on seedlings of the common Sophora about 6 to 8 feet high ; and from this elevation the branches hang down until on reaching the ground their tips spread out or turn up.   It can be used as an arbour; and even in winter the light, smooth, green branches make it ornamental.   The only trouble is in procuring smooth, straight stems of the ordinary Sophora of a sufficient height.   F. L. Temple[3] says : " In spring plant dormant Sophoras about ¾ inch in diameter in the fairly rich earth bottom of a greenhouse.   Cut them back to the ground, and set them 1 foot apart each way ; and by December first they will be out of the top of the house and as smooth as willows.   Then lift and keep them protected in a cellar or frame, or heel them deep in a well-drained place till spring, when they can be planted in nursery rows, and grafted at the same time with the most gratifying results."   With regard to the origin of the weeping Sophora nothing is known definitely ; but Fortune[4] states that at Shanghai in 1853 he saw "pretty specimens of *Sophora japonica pendula*, grafted high as we see the weeping ash in England." It is probable that this variety was imported early from China.

Var. *crispa.*—Leaves curled, the points of the shoots resembling as it were clusters of ringlets.   We have never seen a specimen of this curious variety, which is not mentioned in the Kew Hand-list.

Var. *Korolkowii* (*Sophora Korolkowii*, Cornu).[5]—This has longer and narrower leaflets than the type, and the young shoots, leaf-stalk petiole, and under surface of the leaflets, are whitish pubescent.   The flowers are said to be of a dirty white in colour.   Köhne[6] states that one of Dieck's introductions from Mongolia is identical with the plant cultivated at Segrez under this name, the origin of which is unknown. In the summer of 1904 I visited the Arboretum at Segrez, and saw this

---

[1] A Plate showing buds will appear in a later part.

[2] An excellent article upon different species of weeping trees was published in the *Gardeners' Chronicle*, 1900, xxviii. 477 ; and on p. 479 there is a good figure of a fine specimen of the weeping Sophora.

[3] *Garden and Forest*, 1889, 164.   [4] Fortune, *Residence among the Chinese*, 139.

[5] Cornu's name is given on the authority of Zabel, in *Laubholz-benennung* (1903), p. 256.   We have been unable to find Cornu's description of the species.   [6] Köhne, *Dendrologie*, 1893, p. 323.

specimen, which is about 30 feet high with a stem a foot in diameter, bearing a large roundish crown like the common Sophora. In the absence of flowers or fruit, it is impossible to say whether it is a distinct species; but in foliage and other characters it differs so little from *Sophora japonica*, that probably Zabel[1] is correct in considering it to be only a form of that species. It seems to be well worth cultivation, judging from the vigorous growth and dense foliage of the fine specimen at Segrez.

Var. *violacea.* — This variety has also whitish pubescence on the shoots, petiole, and under surfaces of the leaflets, which are longish, with an acute or acuminate apex. The flowers are violet according to Dieck.[2] It does not appear to be in cultivation in England.

I incline to the belief that we have in these forms to deal with only two varieties of *Sophora japonica*, which is a widely spread species, and presents considerable variation in pubescence and in colour of the flowers in China.

Var. *oligophylla*, Franchet.[3]—This is a curious variety found by Père David at a tomb near Peking, where he observed two trees. The leaflets are very few in number, three or four, and the end one is trilobed; they are thicker in texture and more glaucous than is ordinarily the case. This variety would be well worth introduction.

DISTRIBUTION AND HISTORY

Sophora japonica, in spite of its name, does not appear to be really wild in Japan, although it is recorded from that country by Franchet[4] and Matsumura.[5] Shirasawa,[6] the latest Japanese authority, says it is planted around habitations in both the sub-tropical and temperate regions of Japan, and that it was introduced from China. Sargent[7] observes: "Even Rein (*The Industries of Japan*), usually a most careful observer, states that '*Sophora japonica* is scattered throughout the entire country, especially in the foliaceous forests of the north.' He had evidently confounded Sophora with Maackia,[8] a common and widely spread tree, especially in Yezo. Sophora, which is only seen occasionally in gardens, does not appear to be a particularly popular tree with the Japanese." The Kew Herbarium specimens from Japan are from gardens near Nagasaki, no wild specimens having been ever received.

Sophora japonica is undoubtedly a native of China, and it is recorded from nearly all the provinces where Europeans have made botanical collections; but of its occurrence as a forest tree there is little information. It appears to be really wild in the province of Chihli. I have never seen it in the numerous mountain forests which I visited in Central China or Yunnan; and it is difficult to decide whether the trees seen at lower levels, where cultivation has been going on for centuries, are wild or planted. It has a wide range as a cultivated plant in China, as it flourishes from Pekin to Hongkong and from Shanghai to Yunnan.

---

[1] See note 5 *supra*.   [2] Köhne, *loc. cit.*   [3] Franchet, *Plantæ Davidianæ*, i. 100 (1884).
[4] Franchet et Savatier, *Enum. Plant. in Japonia*, i. 115.   [5] Matsumura, *Shokubutsu-mei-i*, 279 (1895).
[6] Shirasawa, *loc. cit.*, Text, i. 86. The tree is called *Enju* in Japan.   [7] Sargent, *Forest Flora of Japan*, 1.
[8] Maackia is another name for *Cladrastis amurensis*; but it is possible that the tree confused with *Sophora japonica* in Japan is *Sophora platycarpa*, Maxim, which is very similar to it in foliage. Dupont, *Les Essences Forestières du Japon*, p. 66, gives a very complete account of the wood and the uses of a forest tree in Japan, which he considered to be *Sophora japonica*; but as it is evidently not that species, and as it is uncertain whether he referred to *Sophora platycarpa* or *Cladrastis amurensis*, I have not quoted his description.

It has been known to the Chinese from the earliest times, and has been always named by them the *Huai* tree. In the *Chou Li*, a Chinese classical book, dating from several centuries before the Christian era, it is mentioned as having a place in official audiences. In front of the high officials were placed three Sophora trees, beside which stood the counsellors. It was also used as firewood, and was planted in cemeteries, being the tree prescribed by law to be planted beside the tumulus, 4 feet high, in which officials of low degree were buried. The tumulus of the emperor was 30 feet high, and around it pine trees were planted. Feudal princes were honoured with cypresses; and common people were only permitted to have willows around their tombs. The Sophora was also used medicinally from the earliest times in China, the flowers, fruit, bark, and root being all employed. In the *Erh-ya*, the oldest Chinese dictionary (twelfth century B.C.), the Huai tree is called the *guardian of the palace*; and it is said to open its leaves by night and close them by day. The text is probably corrupt, and the periods of expanding and folding of the leaves are reversed. This is perhaps the first reference in any literature to the phenomenon of the sleep of plants. The term *guardian of the palace* no doubt refers to its use in official audiences.

With regard to the uses of Sophora in China at the present day, in addition to its ornamental character as a tree planted frequently in the courtyards of temples, it is also of considerable economic importance. In commerce the flower-buds (*Huai-mi, huai-hua, huai-tze*), and pods (*huai-chio, huai-shih*) are met with everywhere; but considerable confusion has arisen in books as to the exact uses of these products. Shirasawa (*l.c.*) is inaccurate in stating that the Chinese use the bark to dye paper and cloth of a yellow colour. Mouillefert[1] says the leaves are used for dyeing; but this is also an error. The facts are simple: the flower-buds are used as a dye, and the pods as a medicine.

The flower-buds, as seen for sale, are mixed with stalks, etc., and are evidently collected when quite young as they are only about $\frac{1}{8}$ to $\frac{1}{4}$ inch long. They are oval and pointed at the stalked end, dark greyish in colour, and tasteless. When immersed in water they impart to it a fine yellow colour. These flower-buds, packed in large sacks, are exported in considerable quantity from Shanghai and Tientsin. Consul Meadows in a letter to Kew gives an account of the process of dyeing, which is one for dyeing blue cloth a green colour rather than for obtaining a yellow colour.[2] Debeaux[3] asserts that the buds are moistened with water, and a quantity of common salt is added; the mixture is then put in a press, which squeezes out a liquor with which cotton or silk may be dyed yellow. He adds that the leaves do not contain any yellow colouring matter.

Every part of the tree abounds in a purgative principle; and it has been asserted that it is dangerous to work with the wood when it is fresh, owing to the

---

[1] Mouillefert, *Traité des Arbres*, 629.

[2] The process, according to Meadows, is as follows :—" To dye a piece of cotton cloth of narrow width ($1\frac{1}{2}$ feet) a thousand feet long, a mixture is made of 42 lbs. of Sophora buds, 8 lbs. of alum, and 666 lbs. of water, which is boiled in a large pot for six hours. In Chekiang both cottons and silks are first dyed a light blue, and are then put in the mixture just described, and all is boiled over again for three or four hours; the cloth is then taken out and dried in the sun. It is afterwards boiled and sun-dried once or twice again, according as a lighter or darker tint of green is required."

[3] Debeaux, *Note sur quelques matières tinctoriales des chinois* (1866).

distressing symptoms which ensue; and turners of the wood suffer especially. The active principle resembles the *cathartine* which occurs in senna leaves. In the botanical garden at Dijon there is a well beneath a Sophora tree, and when its leaves or flowers are about to fall the gardener covers the well, having found by experience that the water acquires laxative properties by the infusion in it of the Sophora leaves or flowers.[1]

The wood, according to Shirasawa (*l.c.*), differs remarkably in the colour of the heart-wood and sap-wood; the specific gravity is in dry air 0.74. It is tough and durable, though light and coarse grained; and the annular layers are marked by broad bands of open cells. In Japan it is used for the pillars and frames of their wooden houses, but is not of sufficient importance to have been included in the Japanese Forestry exhibit at St. Louis, nor is it mentioned in Goto's *Handbook of the Forestry of Japan* as a valuable wood.

INTRODUCTION

Petiver[2] (1703 or a little earlier) speaks of " Hai-hoa, *Chinensibus, flore albo, siliquis gummosis articulatis*," evidently the Sophora, and it is probable that the specimen was collected in the island of Chusan by Cunningham in 1700.

Desfontaines,[3] quoting Guerrapain,[4] states that the tree was first raised in Europe from seeds sent by Père d'Incarville (a Jesuit stationed at Peking) in 1747, the first trees being planted at the Petit Trianon by B. de Jussieu. It was unknown to what genus the tree belonged, until it flowered near Paris in 1779. It was introduced in 1753 into England by James Gordon, a celebrated nurseryman at Mile End.[5] Mr. Nicholson obtained from Mr. James Smith, former curator of Kew Gardens, some interesting details concerning the Kew trees. Five plants were early planted at Kew, all of which were still there in 1864, but two no longer exist. One of the three trees remaining is near the rockery; not far off is the famous specimen in chains, while the third tree is in the village at Kew beside the house once occupied by Mr. Aiton, the first director of the Kew Gardens. These three trees, according to Mr. Nicholson,[6] are probably as old as any existing elsewhere in England. There is, however, another tree at Kew beyond the Pagoda of which there is no history.

CULTIVATION

Sophora japonica is an ornamental tree, the peculiarities of which make it interesting. The leaves are dark, glossy green, of an unusual tint, and the younger branchlets are of the same colour. The leaves fall very late in autumn, and keep on

---

[1] Loudon (ii. 564), quoting from Duhamel, states that the bark and green wood of this tree exhales a strong odour which produces on those who prune it a remarkable effect. A plank cut from a tree at Kew in Elwes' possession shows a hard, compact, yellowish brown wood.

[2] Petiver, *Musei Petiveriani Centuriæ decem rariora Naturæ continens*, No. 930 (1692-1703).

[3] Desfontaines, *Histoire des Arbres*, ii. 258 (1809). [4] Guerrapain, *Notice sur la culture du Sophora*.

[5] *Hort. Kew*, first edition (1789), ii. 45. In *Andrews Repository*, ix. 585, there is a figure of a specimen from a tree 40 feet high in the collection of John Ord at Purser's Cross, Fulham, which was planted by him forty years before. Ord is stated to have received his plants from Gordon, " who introduced the species from China in 1753." It is also stated that the Sophora first flowered in England at Syon in August 1797. Loudon, however (*loc. cit.*), states that " the oldest tree near London is at Purser's Cross, where it flowered for the first time in England in August 1807."

[6] Nicholson in *Woods and Forests* (1884), p. 500.

the tree fresh and green long after most trees have lost their foliage. The time of flowering is also very late, and this is a point of interest, although the flowers are not conspicuous or remarkable for size or colour. It is a very hardy tree in England,[1] and seems to be free from all attacks of fungi and insects. Its roots do not sucker, which is a point in its favour when planted in towns or in gardens or parks. It has been freely used as a street tree in Italy, where its dense foliage is an advantage in the hot summers. It is remarkable how little the foliage is affected by the hottest and driest seasons, and on this account it might be tried in dry and hot situations. It thrives fairly well in all soils that are deep and not too compact, but it will only grow vigorously in deep rich soils, where seedlings will sometimes attain a height of 12 feet in four or five years.[2] It is propagated by seeds, which should be sown in spring.

REMARKABLE TREES

The trees in the Kew Gardens have been alluded to as regards their history. The one which occurs near the Pagoda, in 1903, was 68 feet high and 8 feet 3 inches in circumference. The old tree, with the branches held together by chains, now measures (1905) 50 feet high and 13 feet in girth at a foot from the ground, the narrowest part of the short bole, which branches immediately into three main limbs. A fourth limb, very large, was blown off some years ago. Not far off is a smaller tree about 6 feet in girth near the ground; it branches from the base, forming a wide-spreading low tree.

At Syon, two trees of considerable size are now living, each about 70 feet high; one measured in 1903 $12\frac{1}{2}$ feet in girth, the other 12 feet.[3]

The tree in the Oxford Botanic Garden was 65 feet high by 12 feet 3 inches in girth in 1903 when measured by Elwes.[4]

That in the old Botanic Garden at Cambridge is one of the finest trees in England, as it has a very symmetrical bole. It measured in 1904, 73 feet high by 11 feet in girth. It is figured in Plate 16.[5]

We are not acquainted with any large specimens of the Sophora now growing in Scotland, Ireland, or Wales, though Loudon mentions one at Tyninghame, Haddingtonshire, 42 feet high,[6] one at Castletown near Dublin 35 feet high, and one at Oriel Temple, Co. Louth, of the same height.

In France and Germany there are probably larger specimens than in this country.   (A. H.)

---

[1] The Sophora has withstood, without injury, the severest frosts in Perthshire. See pamphlet by Col. H. M. Drummond Hay, *The Comparative Hardihood of Hardwooded Plants, from Observations made at Seggieden, Perthshire* (1882).

[2] Nicholson, in an excellent article on the Sophoras in *Woods and Forests*, July 30, 1884.

[3] One of these is mentioned by Loudon, ii. 565, as being the largest near London, and measured in 1838 57 feet high and about 9 feet in girth.

[4] This is said by Loudon (*l.c.*) to have been twenty years planted in 1844, though probably this is an error, as it was then 35 feet high.

[5] Loudon says there were two trees in the garden, both 50 feet high, which had flowered occasionally.

[6] There is a splendid Sophora in the grounds at Cobham Park, Kent, which I measured in 1905, and found to be 85 feet by 10 feet. There is also one in the Tilt Yard at Arundel Castle, 62 feet by 9 feet 6 inches.—(H. J. E.)

ARAUCARIA

Araucaria, Jussieu, *Gen. Pl.* 413 (1789); Bentham et Hooker, *Gen. Pl.* iii. 437 (1880); Masters,
  Journ. Linn. Soc. (*Bot.*) xxx. 26 (1893).
Dombeya, Lamarck, *Dict.* ii. 301 (*non* Cavanilles) (1786).

TALL evergreen trees, with naked buds and coriaceous leaves, which are widest
at their bases and spirally arranged on the shoots.[1] Usually diœcious. Male flowers
in catkin-like masses, solitary or in fascicles at the ends of the branchlets; anthers
numerous, with a prolonged connective, from which hang six to fifteen pollen sacs.
Female flowers terminal, composed of many scales spirally arranged in a continuous
series with the leaves, there being no obvious distinction between the seed-scale and
the bract; each scale bears one ovule attached to the scale along its whole length.
Cones globular, composed of imbricated wedge-shaped scales thickened at the apex.
Seeds, one on each scale and adnate to it, flattened and without wings.

The genera Araucaria and Agathis constitute the tribe Araucarineæ, which are
distinguished from the other Coniferæ by having a single ovule on a simple scale.
In Agathis the ovule is free from the scale, while in Araucaria it is united with it.
Cunninghamia, which was considered by Bentham and Hooker and by Masters to
belong to this tribe, is now generally classed with the Taxodineæ; in it each scale
bears three ovules.

There are about ten species of Araucaria, inhabitants of South America, Australia,
New Guinea, New Hebrides, New Caledonia, and Norfolk Island. *Araucaria
Cunninghami* has been reported several times as growing in the open air in England;
but in some cases it is evident that *Cunninghamia sinensis* was the tree in question,
while in other cases small plants were referred to which were speedily killed by the
cold of our winters.[2] *Araucaria imbricata* is the only species which is hardy in this
country. There are fine specimens of some of the other species in the Temperate
House at Kew, viz. *Araucaria Bidwilli*, 48 feet high; *Araucaria excelsa*, 48 feet;
Araucaria Cunninghami, 47 feet; and *Araucaria Cookii*, 30 feet.

---

[1] *Araucaria Bidwilli* has the leaves also spirally arranged, but by twisting on their bases they assume a pseudo-
distichous appearance.

[2] In a letter in the *Gardeners' Chronicle*, May 1, 1869, Mr James Barnes, then gardener at Bicton, states, in reply to a
suggestion that the tree there might be *Cunninghamia*, that it was really *Araucaria Cunninghami*, and that it had attained a
height of 36 feet, with a diameter of branches of 28 feet, in a sheltered plantation in that favourable locality. But this tree
was no longer living when I visited Bicton in 1902.—(H. J. E.)

ARAUCARIA IMBRICATA, Chilean Araucaria

Araucaria imbricata, Pavon, in *Mem. Acad. Med. Madrid*, i. 199 (1797); Lambert, *Genus Pinus*,
  106, t. 56, 57 (1832); Loudon, *Arb. et. Frut. Brit*. iv. 2432 (1844); Kent in *Veitch's Man.
  Coniferæ*, ed. 2, 297 (1900).
Araucaria Dombeyi, A. Rich. *Conif*. 86, t. 20 (1826).
Araucaria chilensis, Mirb., *Mem. Mus. Par.* xiii. 49 (1825).
Araucaria araucana, C. Koch, *Dendr*. ii. 206 (1873).
Pinus araucana, Molina, *Sagg. Storia Nat. Chile*, 182 (1782).
Dombeya chilensis, Lamarck, *Encycl.* ii. 301 (1786).

Araucaria imbricata is the oldest name under the correct genus Araucaria, and is, moreover, the one
most generally used.   *Piñon* is the Spanish name in Chile, *Pehuen* the Indian name.

Araucaria imbricata is a tree usually 50 to 100 feet high,[1] with a cylindrical stem,
only slightly tapering in adult trees, and attaining 3 to 5 feet in diameter.   The
bark is very rough and divided into large thick irregularly pentagonal or hexagonal
scales.   The branches, in whorls of 6 or 7, are at first very spreading, and in
young or isolated individuals persist for a long time, but in the forest generally fall
off until a broad umbrella-shaped crown of very crowded branches remains.   In
certain cases,[2] secondary shoots appear on the trunk among the older branches as
they die off.

Leaves: all of one kind, spirally crowded on the branches, sessile, coriaceous,
rigid, ovate-lanceolate, with a sharp point at the apex, slightly concave on the upper
surface, glabrous, bright shining green, marked with longitudinal lines, bearing
stomata on both surfaces, margins cartilaginous; persistent for 10 or 15 years,
withering during the later period of their life; their remains may be seen for a long
time on the trunk and branches as narrow transverse ridges.

Male flowers: catkins almost cylindrical in shape, solitary or 2 to 6 in a
cluster, terminal, sessile, erect, 3 to 5 inches long, yellow in colour, composed of
densely packed anther scales, the tips of which are sharply pointed and recurved;
pollen sacs 6 to 9.   The male flowers frequently remain intact on the tree for
several years; they generally in Europe appear early in spring, the pollen escaping
in June or July.

Female flowers: ovoid, solitary, terminal, erect, about 3 inches long, composed
of numerous wedge-shaped scales, terminating in long, narrow, brittle points.

Cones: globular, brown in colour, 4 to 6 inches in diameter, falling to pieces
when the seeds are ripe (in England in late summer, in Chile in January or
February).   The cones take two years to ripen, fertilisation occurring in the second
year in June or July, when the scales open and expose the ovule to the pollen blown
from neighbouring staminate trees.   Three months after fertilisation the seeds are
fully matured.

Seeds: adnate to the scale and falling with it, 1 to 1½ inch long, wingless,
covered by a thick brown coat.   There are about 300 seeds in a cone.

---

[1] I have seen in Chile trees exceeding even 100 feet in height.—(H. J. E.)
[2] Such a case exists in a large tree at Tortworth Court.—(H. J. E.)

Seedling.—The cotyledons are two in number, and on germination remain below the soil enclosed in the seed (Plate 15 F). The caulicle, to which is attached the cotyledons, is thick, fleshy, and carrot-shaped, serving as a store of nutriment for the plant after that of the cotyledons is exhausted; it is directed downward into the soil, and terminates in a long, slender, fibrous root, which gives off a few lateral rootlets. The plumule, the portion of the axis with its accompanying leaves, which is formed in the embryo prior to generation, protrudes between the stalks of the cotyledons, speedily becomes erect, and develops into the young stem, which bears leaves similar in shape to those of the adult plant. The cotyledons sometime after the stem has grown well above ground wither away, the ends of their stalks being visible on the upper part of the caulicle. At the end of the first season the stem is 4 or 5 inches long, and bears alternate leaves about ¾ inch long, gradually increasing in size from below upwards and forming a crowded tuft at the summit. The lower end or so of the stem is reddish, with leaves small and scale-like. The fusiform caulicle, about an inch in length, is continued below into a root 8 or 9 inches long.

Sexes.—The Araucaria is usually diœcious, the trees being either male or female. It was long supposed that there was a difference in the habit of the two sexes, due, doubtless, to Pavon's account of the matter. Araucarias differ, however, remarkably in habit, and no inference can be drawn as to sex from the habit or character of the growth of an individual. Monœcious trees (as is the case in nearly every diœcious species) are of exceptional and very rare occurrence. The most noted of these occurred at Bicton.[1] Other cases have been recorded from South Lytchett,[2] near Poole, and Pencarrow in Cornwall.[3]   (A. H.)

DISTRIBUTION

This remarkable tree was discovered in or about 1780 by a Spaniard, Don Francisco Dendariarena, who was employed by the Spanish Government to examine the trees in the country of the Araucanos, with the object of finding out those whose timber was best suited for shipbuilding. His account of its discovery, as quoted by Lambert, pp. 106-108, is as follows :—

"In September 1782 I left my companion, Don Hippolito Ruiz, and visited the mountains named Caramavida and Nahuelbuta belonging to the Llanista, Peguen, and Araucano Indians. Amongst many plants which were the result of my two months' excursion, I found in flower and fruit the tree I am about to describe.

"The chain or cordillera of the Andes offers to the view in general a rocky soil, in parts wet and boggy, on account of the abundance of rain and snow which fall in these regions, similar to many provinces in Spain. There are to be seen large forests of this tree which rises to the amazing height of 150 feet, its trunk quite straight and without knots, ending in a pyramid formed of horizontal branches which decrease in length gradually towards the top, and is covered with a double bark, the

---

[1] *Gard. Chron.* 1890, viii. 588, 593, Fig. 118.   [2] *L.c.* 753.   [3] Specimens in the museum at Kew.

inner 5 or 6 inches thick, fungous, tenacious, porous, and light, from which as from almost all other parts flows resin in abundance ; the outer is of nearly equal thickness, resembling cork cleft in various directions, and equally resinous with the inner."

I may say that the district spoken of is not really part of the Andes at all, but a coast range separated from the Andes by a wide tract of low country, mostly covered with forest. And as regards the bark, though I did not see any old trees felled in Chile, the bark of trees of 40-50 years old felled in England does not show bark at all approaching the thickness described. Neither have I seen in the districts I visited myself any trees as tall as he describes, or more than about 120 feet. He states that it is also found "juxta oppidum Conceptionis." There are no mountains near Concepcion high enough for the Araucaria, and I think this must be based on false information.

Don Dendariarena goes on to say that "the wood of this tree is of a yellowish white, fibrous, and full of very beautiful veins, capable of being polished and worked with facility. It is probably the best adapted for shipbuilding, as has been shown by the experiments made in the year 1780, in consequence of which orders were given to supply the squadron commanded by Don Antonio Bacaro, then at anchor in the port of Talcahuano."

"The resin abounding in all parts of the tree is white, its smell like that of frankincense, its taste not unpleasant. It is applied in plaster as a powerful remedy for contusions and putrid ulcers, it cicatrises recent wounds, mitigates headaches, and is used as a diuretic, in pills, to facilitate and cleanse venereal ulcers. The Indians make use of the fruit of this tree as a very nourishing food ; they eat it raw as well as boiled and roasted, with it they form pastry, and distil from it a spirituous liquor."

Lambert says : "In a letter which I have lately received M. Pavon mentions an important particular, not noticed in the above description, namely, that the male tree is not above half the size of the female, and seldom exceeding 40 feet in height." I am not able to confirm this from personal observation either in Chile or England, and Dr. Masters[1] says that there is no reliable distinction between the male and female tree, whilst it is said in an account of the Araucarias in the Piltdown Nurseries[2] that the habit of the tree is no guide to the sex.

It was first described by the Abbé Molina, who called it *Pinus araucana.* Ruiz and Pavon who explored parts of Chile soon afterwards sent specimens to Europe to a Frenchman named Dombey, which were described by Lamarck under the name of *Dombeya chilensis,* but the generic name he gave cannot stand because it was previously used for a genus of Sterculiaceæ.

In 1795 Captain Vancouver visited the coast of Chile, accompanied by Archibald Menzies, who procured some seeds which he sowed on board ship,[3] and succeeded in bringing home living plants, which he gave to Sir Joseph Banks, who planted one of them in his own garden at Spring Grove, and sent the remaining five plants to Kew. One of these, after being kept in the greenhouse till about 1806 or 1808,

---

[1] *Gard. Chron.* 1890, ii. 667.   [2] *Ibid.* 1891, i. 342.

[3] Sir Joseph Hooker, who knew Menzies personally, tells me that he took these seeds from the dessert table of the Governor.

was planted out on what is now called Lawn L, and was at first protected during winter by a frame covered with mats. Here it grew for many years and attained the height of 12 feet in 1836 (*fide* Loudon), but eventually died in the autumn of 1892 at the age of nearly 100 years.[1] This is probably the tree figured by Lambert.

The first person who gives any account of the tree in its native forests, so far as I know, is Dr. Poeppig, whose account of the tree is printed in *Companion to Bot. Mag.* i. 351-355. It did not, however, become common in cultivation till the celebrated botanical traveller William Lobb, who was sent to South America by the firm of Veitch, sent home in 1844 a good supply of seeds which produced most of the finest trees now in England.

No account of his travels were, however, published, and on applying to Messrs Veitch before I went to Chile in 1901 I was informed that his journals, which I wished to consult, could not be found. The late Miss Marianne North was the first English traveller who published any account of the tree in its native forests, which she visited on her last journey in November 1884, mainly, as she says, for the purpose of painting this tree. But, owing to the difficulty and danger at that time of reaching the Andes, she went to the coast range of Araucania, called Nahuelbuta, which lies between the sea and the town of Angol, in the same district where the tree was probably first discovered. After describing her ride up from Angol to the mountains, which are here covered with a beautiful vegetation, among which Gunnera, Lapageria, Embothrium, Fuchsia, Buddleia, Alstroemeria, and many other favourite plants in English gardens are conspicuous, she says:[2]—

"The first Araucarias we reached were in a boggy valley, but they also grew to the very tops of the rocky hills, and seemed to drive all other trees away, covering many miles of hill and valley; but few specimens were to be found outside that forest. The ground underneath was gay with purple and pink everlasting peas, and some blue and white ones I had never seen in gardens, gorgeous orange orchids, and many tiny flowers whose names I did not know, which died as soon as they were picked, and could not be kept to paint. I saw none of the trees over 100 feet in height or 20 in circumference, and, strange to say, they seemed all to be very old or very young. I saw none of the noble specimens of middle age we have in English parks, with their lower branches resting on the ground. They did not become quite flat at the top, like those of Brazil, but were slightly domed like those in Queensland, and their shiny leaves glittered in the sunshine, while their trunks and branches were hung with white lichen, and the latter weighed down with cones as big as one's head. The smaller cones of the male trees were shaking off clouds of golden pollen, and were full of small grubs; these attracted flights of bronzy green parrakeets, which were busy over them. Those birds are said to be so clever that they can find a soft place in the great shell of the cone when ripe, into which they get the point of their sharp beak, and fidget with it until the whole cone cracks and the nuts fall to the ground. Men eat the nuts too, when properly cooked, like chestnuts. The most remarkable thing about the tree is its bark,

---

[1] Cf. *Kew Bull.* 1893, p. 24.
[2] Marianne North, *Recollections of a Happy Life*, 2nd ed. ii. 323, 324 (1892).

which is a perfect child's puzzle of slabs of different sizes, with 5 or 6 distinct sides to each, all fitted together with the neatness of a honeycomb. I tried in vain to find some system on which it was arranged. We had the good fortune to see a group of guanacos feeding quietly under the old trees. They looked strange enough to be in character with them, having the body of a sheep and the head of a camel; and they let us come quite near. On the other side of the mountains they are used as a beast of burden, though so weak that ten of them could not carry the load of an average donkey. After wandering about the lower lands, we climbed through the bogs and granite boulders to the top of one of the hills, and came suddenly to a most wonderful view, with seven snowy cones of the Cordillera piercing their way through the long line of mist which hid the nearer connecting mountains from sight, and glittering against the greenish blue sky. Each one looked perfectly separate and gigantic, though the highest was only 10,000 feet above the sea. Under the mist were hills of beech forest, and nearer still the Araucaria domes, while the foreground consisted of noble old specimens of the same trees grouped round a huge grey boulder covered with moss and enriched with sprays of embothrium of the brightest scarlet. No subject could have been finer, if I could only have painted it, but that 'if' has been plaguing me for years, and every year seems to take me farther from a satisfactory result."

Inspired by this charming description, and by a desire to see the magnificent forests of Southern Chile, whence I hoped to introduce new trees and plants to our gardens, I visited Chile in the winter of 1901-1902, and after various difficulties caused by the dispute about the frontier, which nearly led to a war between Chile and Argentina, I started from the hospitable home of my friends, Mr. George and Senora Bussey at San Ignacio, to see the Araucarias in the Sierra de Pemehue, a region where they attain their greatest perfection, and which, having only been recently conquered from the Indians, had been described by no scientific traveller; though Senor Moreno has written an excellent account of the Argentine side of the frontier, which I visited later.

The Sierra de Pemehue is a range of mountains lying on the west side of the upper course of the great Bíobio river, and is not, strictly speaking, a part of the Cordillera of the Andes, from which it is separated by that river. The greater part of it is covered with splendid forests, principally composed of beeches, *Fagus obliqua* and *Fagus Dombeyi*, and it was near the head-waters of the Renaico river that I first saw what is to me the most striking of all trees hardy in England, and the only Chilean tree which as yet seems to have acclimatised itself thoroughly in this country.

They were growing in scattered groups on the cliffs far above us at an elevation of 3000-4000 feet, and we did not enter the Araucaria forest till we got near the top of the pass, which crossed over a mountain called Chilpa, between the Renaico and the Villacura valleys. Here the trees were growing scattered among Coigue trees (*Fagus Dombeyi*), and higher up in a forest mainly composed of Niere (*Fagus antarctica*), many of which were killed by forest fires, which had not, however, destroyed the thick-barked Araucarias, though I saw here but few young trees and no seedlings.

Their average height was 80-90 feet, and the diameter 2-3 feet, and the branches were mostly confined to the top of the tree, where they form a dense, flat-topped crown. On 27th January I saw much finer specimens in the valley above Lolco, on the road to Longuimay, and my companion, Mr. Bartlett Calvert, was successful in getting some excellent photographs which are here reproduced. Plate 17 shows the appearance of mature and young trees growing in an open grassy valley at about 4500 feet, with the high volcanoes of Longuimay and Tolhuaca in the background. The old tree on the right of the picture is about 90 feet, and the young one about 20 feet high, showing sixteen years of growth from a point 2-3 feet from the ground where the annual growths could no longer be distinguished. I therefore suppose this young tree to be twenty to twenty-five years old from seed.

Farther on in the same valley we came to much larger trees, which showed the curiously irregular slabs of bark of which Miss North speaks. The largest trees I saw had a girth of 24 feet at breast height, and were 90-100 feet high. The longest fallen stems I measured were little over 100 feet, and I should say 80-90 was the average height of full-grown ones. Plate 18 shows the habit which the trees assume when grown thickly at about 3500 feet elevation in the upper Villacura valley.

On the wind-swept ridges which we crossed higher up the pass, at an approximate elevation of 6500 feet, the Araucarias were much more stunted and had a very different habit of growth, but the high wind which prevailed, as it usually does at this season, made it impossible to photograph them. Two days later at Los Arcos, the frontier post of Argentina, I found scattered groves of Araucaria for about fifty miles south, as far as the valley of Quillen, but when we reached the country about the head-waters of the Pichelifeu river, about lat. 39° 30′ S., I saw no more except a few isolated trees which appeared to have sprung up from seeds dropped by the Indians on their old camping grounds.

I had previously been told by Mr. Barton of Buenos Ayres, who is engaged in cutting timber on the north shore of the great lake Nahuelhuapi, about 100 miles to the south, that the Araucaria was found near this lake, and I had great hopes of discovering and introducing a new southern variety or species, which might prove hardier than *A. imbricata*.

But notwithstanding what Poeppig says as to the probability of its extension as far south as lat. 46°, I saw not a single tree on my journey from San Martin *via* Nahuelhuapi to Puerto Montt in lat. 41° 50′, and none of the explorers who have been recently employed in surveying the frontier have, so far as I know, found it south of about lat. 41°. Sir T. Holdich is my authority for this statement.

Some of the trees here had much smoother bark covered with long tufts of grey lichen, and in this part of the forest there were plenty of young seedlings coming up, some of which I took up and unsuccessfully attempted to transplant to my friend's garden at San Ignacio.

The geographical range of the tree is therefore a very limited one, extending only from Antuco in about lat. 38° 40′ to lat. 40° in the Cordillera, and on the coast range from about lat. 38° 30′ to an unknown point probably not south of about lat. 41°. For, though Poeppig says it occurs on the Corcovado, he was speaking only from hearsay, and the sudden change of the climate, which here becomes an extremely wet one, is probably the reason why the tree does not exist on the west coast in a much higher latitude, as do the majority of the trees and plants which are associated with it.

Another point in which I must differ from Poeppig is the bareness of the Araucaria forests of other vegetation. Though, of course, where the trees are closely crowded not many plants grow in their shade, yet the number of beautiful terrestrial orchids and other plants which I found in the more open parts of the Araucaria forest was very striking, and Miss North's observations in the Nahuelbuta range quite confirm my opinion that the moderate shade of the Araucaria is not prejudicial to herbaceous plants.

The soil on which it grows is mostly of volcanic origin, sometimes covered with deep vegetable mould, but more usually dry and rocky; and the climate, though warm and dry in the months of December, January, and February, is cold and wet in winter.

The only exact particulars I can give of the climatic variations were taken during the winter of 1901 at Rahue in the upper Biobio Valley, near Longuimay, at an elevation of 700 metres, which is lower and thus probably warmer than that of the Araucaria region. These observations I have condensed as follows :—

	MAXIMUM.	MINIMUM CENTIGRADE.	
Between April 21 and 30	+ 26 (on 25th)	− 3 (on 30th)	
„ May 1 „ 31	+ 19 (1st, 25th)	− 7 (on 13th)	snow on 7 days.
„ June 1 „ 30	+ 22 (on 27th)	− 6 (on 24th)	snow on 8 days.
„ July 1 „ 31	+ 12 (on 27th)	5 (on 10th)	snow on 6 days.
„ Aug. 1 „ 30	+ 12 (on 19th)	6 (on 17th)	snow on 5 days. rain on 7 days.
„ Sept. 1 „ 30	+ 24 (on 27th)	? 10 (on 2nd)	snow on 1 day. rain on 12 days.
„ Oct. 1 „ 31	+ 30 (on 17th)	? 6 (on 8th)	snow on 2 days. rain on 5 days.
„ Nov. 1 „ 23	+ 25 (on 21st)	? ?	snow on 2 days. rain on 7 days.

Reduced to Fahrenheit this register shows a very similar climate to that of some parts of England, very variable all the year round, but probably hotter and more sunny in winter.

As regards the summer climate I may say that in the months of January and February, which are the height of summer, it was never cold by day, and the sun and wind often unpleasantly warm, but at night the thermometer often fell to near freezing-point, and on one occasion, on 1st February, my sponge was frozen in camp just south of Lake Aluminé at about 5000 feet. We know that the Araucaria has borne in Great Britain temperatures below zero Fahr. without injury on dry and suitable soil, but it evidently will not endure the continuous wet of the southern coast region of Chile.

In the *Forstliche Naturwissenschaftliche Zeitschrift*, 1897, iv. 416-426, Dr. Neger, who was naturalist on the Chilean Boundary Survey in 1896-97, gives an account of *Araucaria imbricata*, which does not add anything of great importance for English arboriculturists to what I have already stated. He says that there are two types of Araucaria forest, one of which is characteristic of the rainy coast mountain range of Nahuelbuta and the west side of the Andes on the Cordillera of Pemehue; and the other, which is peculiar to the drier plateaux of the Argentine territory, on the east side of the watershed. He refers to Reiche's account of the Nahuelbuta forest in *Engler's Bot. Jahrbuch*, xxii. 110, which gives a good account of the flora. He does not confirm the statement that the male trees are smaller in size than the female, and speaks of trees occurring in deep valleys 40-50 metres high, and 2-2½ metres in diameter at about 3 feet, but does not give any exact measurements, so that this height is probably an estimate by the eye. He says that the seeds do not ripen until May in the year after flowering, but I found them ripe in February and fit to eat in January. He gives some excellent illustrations of Araucaria forests on Nahuelbuta, one of which shows a wider and more unbroken extent than any that I saw; another shows the ability of the tree to take root and grow in the crevices of bare rock. Another shows a forest at the foot of the great volcanic peak of Lanin, where some of the trees have been almost buried by sand and still retain their upright position. Lastly, he gives a small map of the distribution which, however, is not sufficiently detailed to be very accurate; this makes Antuco the most northerly point, and a point somewhere north of lat. 40°, the southerly range of the tree. He says that in the museum of Santiago there are geological evidences of the existence at a former period of Araucaria as far north as the Puna of Atacama.

REMARKABLE TREES

The finest tree which until recently existed in England was at Dropmore, which, however, began to die about four years ago, and was dead when the photograph (Plate 19) was taken in June 1903. It is said[1] to have been purchased at a sale in the Royal Horticultural Gardens at Chiswick in 1829, and in 1893 to have been 69 feet high. When felled in 1905 Mr. Page found it to be 78 feet 6 inches high, and the butt was 27½ inches in diameter at the base under the bark, which was about 2 inches thick, the measurable timber in it being about 65 cubic feet.

There are many fine specimens at Beauport, Sussex, the seat of Sir Archibald Lamb, Bart., where a plantation was made about forty years ago, which gives a better idea of the Araucaria at home than any I have seen in England. It contains 27 trees on an area 102 paces round, and the inside trees are clearing themselves from branches naturally. Twenty of them Sir A. Lamb says are over 50 feet high, and in 1905 I estimated them to contain an average of 25 cubic feet (Plate 20). The largest tree at Beauport, as measured by Henry in 1904, was 74 feet high and 7 feet 9 inches in girth. The trees produce seeds freely, and a seedling growing in a chink of the garden steps was 4 feet high in 1903, and in 1905 had grown at least 2 feet more.

At Strathfieldsaye, Berks, the seat of the Duke of Wellington, the Araucaria has produced self-sown seedlings, a group of which is shown in Plate 15 E.

[1] *Gard. Chron.* 1893, i. 232; also *l.c.* 1872, p. 1324.

At the Piltdown Nurseries in Sussex there are many fine specimens,[1] one of which is said to have been 50 feet by 9½ feet in girth in 1854.   Messrs. Dennett and Sons, the present tenants of this nursery, inform me that they believe this is one of the oldest trees in the country, and that in April 1903 it was about 70 feet high (perhaps more), with a girth of 7 feet at 5 feet, and 11 feet close to the ground.   But a correspondent of the *Gardeners' Chronicle*[2] says that in 1891 it was 65 feet by 10½ feet at 4 feet, and that 3½ bushels of seed were collected in this nursery in 1889, which produced hardier plants than imported seed.   He also states that one of the trees which was cut down in 1880 threw up in 1884 a sucker from the roots, which grew 15 feet high in five years, and showed in 1891 no signs of branching out in any way.[3] He also states that it does not matter when Araucarias are pruned, as they grow steadily all the year.   The soil at Piltdown is a deep loam with gravel subsoil, and though, as it is here stated, it is generally thought that a dry, well-drained subsoil is essential to the success of this tree, yet I have seen in the garden of Foss bridge Inn in the Cotswold Hills, in a low damp situation close to the banks of the Coln, two Araucarias, male and female, about 40 feet high, which produced ripe seed in 1903, from which Mr. Holyoake, gardener to the Earl of Eldon of Stowell Park, has raised plants.

At Bicton, Devonshire, the seat of the Honourable Mark Rolle, there is a fine avenue of Araucarias, which has been often mentioned in print ; but the trees in it do not appear to be increasing in height so fast as the good soil and climate would lead one to expect.   When I saw them in September 1902 the best which I measured was about 50 feet high by 8 feet 9 inches in girth.   Ripe seeds were falling at the time, from which seedlings were raised.

There are also fine trees at Castlehill, North Devon, the seat of the Earl Fortescue, which have produced seed for many years past.

At Tortworth Court, Gloucestershire, the seat of the Earl of Ducie, who has one of the best collections of trees in England, and to whom I am indebted for very much assistance and advice in this work, there are many large Araucarias,[4] the best of which I found to be 53 feet by 7 feet 6 inches in 1904.   It is producing many young shoots among the dying branches of the trunk.

In Scotland the Araucaria grows well not only in the south - west where, at Castle Kennedy, the seat of the Earl of Stair, there is a fine avenue, 200 yards long, in which the largest tree is 50 feet by 6 feet 2 inches in girth, and from which self-sown seedlings have sprung, but also in Perthshire, where there are fair-sized trees, one of which on the banks of the Tay in the grounds of the Duke of Athole at Dunkeld, I found in 1904 to be 50 feet high, but only 3 feet 11 in girth.   It grows well at Gordon Castle exposed to the full force of the north-east wind, and has ripened seeds as far north as Inverness.[5]   But some of the trees recorded in Perthshire and other places in Scotland have been killed during severe frosts, and as a rule the growth is not so rapid as in the south of England.   Two

[1] *Gard. Chron.* 1885, xxiii. 342.   [2] *Ibid.* 1891, i. 342.
[3] Sir Herbert Maxwell informs me that he saw at Cairnsmore an old trunk of Araucaria which had died twenty years ago, still standing, with a young growth 3 feet high from the stool.
[4] Cf. *Gard. Chron.* 1890, ii. 633.   [5] *Ibid.* 1868, p. 464 ; 1872, p. 1323 ; 1894, xvi. 603.

trees at Redcastle, Ross-shire, planted in 1843, measured by Col. A. Thynne, are 47 feet by 7 feet 4 inches, and 40 feet by 6 feet; the latter, though exposed to the east wind, is branched to the ground.

At Ardkinglas there is a very healthy tree 50 feet by 6 feet, and at Inverary, Minard Castle, Poltalloch, and other places in Argyleshire, there are several thriving trees of good size. At Loch Corrie, near Glenquoich, there are two trees at 450 feet above sea-level, one of which in 1905 was 43 feet by 6 feet 2 inches.

In Ireland it seems at home almost everywhere. At Fota, in the extreme south, Henry measured one 62 feet by 5 feet; at Ballenetray, Co. Waterford, a tree was recorded[1] in 1884 as being 65 feet 6 inches by 6 feet; at Woodstock, Co. Kilkenny, there is a tree which in 1904 Henry found to be 65 feet by 9 feet 9 inches; and at Castlewellan, Co. Down, the seat of the Earl Annesley, and many other places, good trees occur.

In the milder parts of Western France the Araucaria thrives, but does not appear to have grown as large as in England. The best is reported by M. de Vilmorin as growing at Penandreff, near St. Renan, Finisterre, which in 1890 was 50 feet high by 7 feet 4 inches in girth. In the *Revue Horticole*, 1899, p. 460, this is confirmed. In Germany I have not heard of any fine examples.

Cultivation and Soil

The Araucaria should always be raised from seed, home-grown seed being preferable; for though plants have been raised from cuttings, which have grown to a considerable size, this mode of propagation is the cause of much disappointment, and of many ill-shaped and unsightly trees, not only in the Araucaria, but in many other conifers. The seed should be sown singly in pots, laying the seed on its side with the thick end in the centre, and will germinate best in a frame or cold greenhouse, where they can be protected from mice and frost. The young plants should not remain in pots more than one or at most two seasons, for though the tap-root does not become so long as in the case of pines, it wants room; and if the climate and soil are not very favourable, the young tree should not be permanently planted out till it is 1 or 2 feet high. The seedlings vary much in vigour, and on cold or calcareous soil many die young; but under better conditions the tree grows at least 1 foot a year when established.

It should be planted only in a well-drained situation, as severe frosts will often kill the trees when small; and though not so particular about the constituents of the soil as most Chilean trees, seems to thrive better on a sandy soil free from lime, especially on the red sandstone and greensand formations.

In the *Gardeners' Chronicle*, August 15, 1885, is an excellent note by Mr. Fowler, whose experience of this tree at Castle Kennedy was extensive, on the cultivation of the Araucaria; and another valuable note on the same subject will be found in the same journal, November 13, 1886, by Mr. C. E. Curtis. Both these authorities consider that the exudation of gum which often occurs in unhealthy trees

---

[1] *Woods and Forests*, Feb. 6, 1884.

is due to the roots of the tree having reached a cold wet subsoil, or from exposure to excessive cold. There seems to be no remedy for this disease, which usually kills the tree.

Araucarias do not thrive in the smoky atmosphere of a large town, and for this reason are not seen at their best in the immediate neighbourhood of London, nor do I know of any very fine ones in Wales or in the midland and northern counties of England.

Uses

The gum which exudes from the bark is used in Chile as a salve for wounds and ulcers. It has a pleasant smell like that of turpentine, and sets hard when dry, but I am not aware that it contains any special intrinsic virtue.

The seeds are largely consumed by the Araucanos and other tribes of Indians, and are occasionally sent for sale to the markets of Valdivia and Concepcion. I have eaten them both roasted and boiled, and found them very palatable, with a nutty flavour somewhat like that of almonds.

The timber is said to have been formerly used in the dockyards of Chile, but is now considered inferior to that of the Alerce (*Fitzroya patagonica*), and perhaps owing to the remote positions in which the trees grow, is not now used except locally. Through the kindness of the Duke of Bedford I received two planks cut from a tree grown at Endsleigh, near Tavistock, of which the wood does not show any specially attractive quality. The Earl of Ducie describes it[1] as " not unlike good deal, but from the absence of turpentine and for some other reason it is smoother to the touch than the ordinary deals of commerce. In this respect its texture is not unlike that of redwood (*Sequoia sempervirens*). On testing a thin batten by breakage, it proved to be tough and strong for its size; but the fracture was abrupt, and showed little longitudinal fibre. The wood is somewhat heavier than ordinary deal." The timber is not mentioned in Stone's *Timbers of Commerce*.   (H. J. E.)

[1] *Gard. Chron.* 1900, ii. 633.

GINKGO

Ginkgo, Linnæus, *Mantissa*, ii. 313 (1771); Bentham et Hooker, *Gen. Pl.* iii. 432, 1225
(1880); Masters, *Jour. Linn. Soc.* (*Bot.*), xxx. 3 (1893).
Salisburia, Smith, *Trans. Linn. Soc.* iii. 330 (1797).

TREES, several extinct and one living species, bearing fan-shaped, fork-veined leaves on both long and short shoots. Flowers diœcious, arising from the apex of short shoots, which bear at the same time ordinary leaves. Male flowers: catkins, 3-6 on one shoot, each being a pendulous axis bearing numerous stamens loosely arranged. Stamen a short stalk ending in a knob, beneath which are 2-4 divergent anthers, dehiscing longitudinally. Female flowers, 1-3, more or less erect on the shoot, each consisting of a long stalk, which bears an ovule on either side below the apex. The ovule is sessile, straight, surrounded at its base by an aril or collar-like rim,[1] and naked (*i.e.* not enclosed in an ovary). Fruit: a drupe-like seed (sessile in the small bowl-shaped little developed aril) consisting of an orange fleshy covering enveloping a woody shell, within which, embedded in the albumen, lies an embryo with 2-3 cotyledons. The albumen is covered by a thin membrane which is only adherent to the woody shell in its lower part. Two embryos often occur in 1 seed, and of the 2 ovules only one is generally developed into a seed.

Ginkgo was formerly considered to belong to the Coniferæ, but recent investigations show that it is distinct from these, and is the type of a Natural order Ginkgoaceæ, which has affinities with Cycads and ferns. The seeds resemble closely those of Cycads, and at the end of the pollen tube are formed two ciliated antherozoids which are morphologically identical with the antherozoids occurring in ferns. Ginkgo, however, is a true flowering plant, as it produces seeds, and is a gymnosperm, since it bears ovules which are not enclosed in an ovary.

The extinct species have been found in the Jurassic and succeeding epochs. Gardner[2] considers the specimens which have been found in the white clay at Ardtun in the Isle of Mull to be specifically identical with *Ginkgo biloba.*

---

[1] Considered now to be a reduced carpel.   [2] J. S. Gardner, *British Eocene Flora* (1886), ii. 100.

GINKGO BILOBA, MAIDENHAIR TREE

Ginkgo biloba, Linnæus, *Mantissa*, ii. 313 (1771); Kent in *Veitch's Man. Coniferæ*, 2nd ed. 107
   (1900); Seward and Gowan, *Ann. Bot.* xiv. 109 (1900).
Salisburia adiantifolia, Smith, *Trans. Linn. Soc.* iii. 330 (1797); Loudon, *Arb. et Frut.*
   Brit. iv. 2094 (1838).

The Ginkgo when young is pyramidal in habit, with slender, upright branches:
older, it becomes much more spreading and broader in the crown. It attains a
height of 100 feet and upwards, with a girth of stem of about 30 feet. Bark: grey,
somewhat rough, becoming fissured when old.

Leaves: deciduous, scattered on the long shoots, crowded at the apex of the
short shoots, which grow slowly from year to year, their older portions being covered
with the leaf-scars of former years. The short shoot may, after several years,
elongate into a long shoot bearing scattered leaves. The leaves are stalked, and
unique in shape amongst trees, recalling on a large scale the pinna of an adiantum
fern; they show much variation in size (2-8 inches in breadth) and in margin,
but generally are bilobed and irregularly crenate or cut in their upper part. There
is no midrib, and the veins, repeatedly forking, are not connected by any cross
veinlets. The stomata are scattered on the lower surface. In the bud the leaves
are folded together and not rolled up, as in the crozier-like vernation of ferns.

Flowers and fruit: see description of the genus.

The drupe-like seeds have a fleshy outer covering of a bright orange colour
when ripe, and when they fall upon the ground, this bursts and emits an odour of
butyric acid which is very disagreeable.[1] They are imperfectly developed as they
fall, though apparently ripe; and the fertilisation of the ovule and the subsequent
development of the embryo occur while they are lying on the ground during winter.
The kernels are edible, being known to the Chinese as *pai-kuo* (white fruits), and are
sold in most market towns of China. They are supposed to promote digestion and
diminish the effects of wine-drinking; and are eaten roasted at feasts and weddings,
the shells being dyed red.

Fruit-bearing trees are now common in Southern Europe; but no fruit, so far
as we know, has ever been produced in England. The well-known tree at Kew is a
male, and produces flowers freely in exceptional years, *e.g.* in 1894, supposed to be due
to the fact that the preceding summer was remarkably warm, with continual sunshine.

Extraordinary cases of abnormal formation of fruit have been observed in Japan.
Shirai[2] described and figured in 1891 fruit which was produced on the surface of
ordinary leaves of the tree. Fujii has studied since then the various stages of the
development of ovules and of pollen sacs upon leaves. The so-called aril of the
fruit is considered by him to represent a carpel, as he has observed transitional stages
between the ordinarily shaped aril and a leafy blade bearing ovules.

---

[1] "The pulp surrounding the seed has a most abominable odour. Although warned not to touch it, I gathered the seeds
with my own hands; but it took me two days' washing to get the odour off."—(W. Falconer in *Garden*, 1890, xxxviii. 602.)
[2] Shirai, in *Tokyo Bot. Mag.* 1891, p. 342.

Jacquin[1] grafted on the male tree at Vienna, when it was quite small, a bud of the female tree, from which a branch developed. This tree is now of large size; and numerous branches regularly bear male flowers, whilst one branch, now very stout, bears female flowers. This female branch puts forth its foliage about fourteen days later than the male branches, and retains them much later in autumn. In this case the shoot retains its individual characters, and the stock does not affect it even in regard to its annual development.

Seedling.—The germination in Ginkgo is not unlike that of the oak. We are indebted to Mr. Lyon[2] of Minneapolis for figures of the seedling, which are reproduced on Plate 15 C, D.

When the seeds are sown the hard shell is cracked at its micropylar end by the swelling of the embryo within. Through this opening the body of the embryo is thrust out by the elongation of the cotyledons, which remain attached to the caulicle by two arching petioles; between these the plumule or young stem ascends, while the root turns down into the soil. The cotyledons remain attached throughout the first season's growth. The first two or three leaves directly above the cotyledons remain small and scale-like. After reaching 4 or 5 inches in height the stem stops growing, having expanded into a rather close crown of ordinary leaves at its apex, which ends in a large terminal bud. The root attains in the first season about the same length as the stem, and develops numerous lateral fibres. This primary root, as is usually the case in Gymnosperms, persists as the tap-root of the plant.

Sexes.—Certain differences, besides those of the flowers, are observable in male and female trees.[3] The male trees are pyramidal and upright in habit, the ascending branches being of free and vigorous growth. The female trees are closer and more compact in habit, more richly branched below, and the branches sometime become even pendent.[4]

Monsieur L. Henry[5] states that in Paris the leaves of the female Ginkgo fell three or four weeks later than those of the male. Generally male trees are completely denuded of foliage by the beginning of November, while the female trees retain their leaves till the end of November or the beginning of December.

Burrs.—In Japan there often develops on old Ginkgo trees peculiar burrs, which are called *chi-chi* or nipples. These may be observed in an incipient stage on the large tree at Kew. They occur on the lower side of the larger branches of the tree, and vary in size from a few inches in length to 6 feet long by 1 foot in diameter. They occur singly or in clusters, and are generally elongated, conical in shape, with a rounded tip. If they reach the ground, as is sometimes the case, they take root, and then bear leaves. They are due to the abnormal development of dormant or adventitious buds. A description of this curious phenomenon and a photograph of a tree bearing a large number of these growths is given by Fujii in *Tokyo Bot. Mag.*

---

[1] Kerner, *Nat. Hist. of Plants* (Eng. trans.) ii. 572.

[2] See Lyon's paper in *Minnesota Botanical Studies*, 1904, p. 275.

[3] Sargent denies this, and says it is impossible to distinguish the sexes till the trees flower; but observations on the Continent go to show that the sexual differences pointed out above really exist. See Sargent, *Garden and Forest*, 1890, p. 549.

[4] See Schneider, *Dendrologische Winterstudien*, 127 (1903), and Max Leichtlin in *Woods and Forests*, Jan. 16, 1884.

[5] *Bull. de l'Assoc. des anc. élèv. de l'école d'Hort. de Versailles*, 1898, p. 597, quoted in *Gard. Chron.* 1899, xxv. 201.

1895, p. 444.   We are indebted to Mrs. Archibald Little for a photograph taken by her in Western China, of a tree 19½ feet round the base, and larger above, which very well shows these excrescences (Plate 23).

IDENTIFICATION

In summer the leaves are unmistakable.   In winter the long and short shoots should be examined.   The long shoot of one year's growth is round, smooth, brownish, and shining, the terminal buds being larger than the scattered lateral buds, which come off at a wide angle.   The buds are conical, and composed of several imbricated brown dotted scales.   The leaf-scars show 2 *small cicatrices*, and are *fringed above with white pubescence*.   The short shoots are spurs of varying length, up to an inch or more, stout, ringed, and bearing at their apex a bud surrounded by several double-dotted leaf-scars.   In Pseudolarix and the larches, which have somewhat similar spurs, the leaf-scars are much smaller, and show on their surface only one tiny cicatrice.   In Taxodium there are no spurs, and the scars which are left where the *twigs* have fallen off show only one central cicatrice.

VARIETIES

The following forms are known in cultivation :—

Var. *variegata*.   Leaves blotched and streaked with pale yellow.

Var. *pendula*.   Branches more or less pendulous.

Var. *macrophylla laciniata*.   Leaves much larger than in the ordinary form, 8 inches or more in width, and divided into 3 to 5 lobes, which are themselves subdivided.

Var. *triloba*.   Scarce worthy of recognition, as the leaves in all Ginkgo trees are exceedingly variable in lobing.

Var. *fastigiata*.   Columnar in shape, the branches being directed almost vertically upwards.[1]

DISTRIBUTION AND HISTORY

The wild habitat of *Ginkgo biloba*, the only species now living, is not known for certain.   The late Mrs. Bishop, in a letter to the *Standard*, Aug. 17, 1899, reported that she had observed it growing wild in Japan, in the great forest northward from Lebungé on Volcano Bay in Yezo, and also in the country at the sources of the great Gold and Min rivers in Western China.   However, all scientific travellers in Japan and the leading Japanese botanists and foresters deny its being indigenous in any part of Japan; and botanical collectors have not observed it truly wild in China.   Consul-General Hosie[2] says it is common in Szechuan, especially in the hills bounding the upper waters of the river Min; but he does not explicitly assert that it is wild there.   Its native habitat has yet to be

---

[1] See *Garden*, 1890, xxxviii. 602.   An interesting article by W. Falconer, who gives some curious details concerning the Ginkgo tree in the United States.

[2] *Parliamentary Papers, China*, No. 5, 1904 ; *Consul-General Hosie's Report*, 18.   Mr. E. H. Wilson in all his explorations of Western China never saw any but cultivated trees.

discovered; and I would suggest the provinces of Hunan, Chekiang, and Anhwei in China as likely to contain it in their as yet unexplored mountain forests.

The earliest mention of the tree in Chinese literature occurs in the *Chung Shu Shu*, a work on agriculture, which dates from the 8th century, A.D. The author of the great Chinese herbal (*Pen-Tsao-Kang-Mu*, 1578 A.D.) does not cite any previous writers, but mentions that it occurs in Kiangnan (the territory south of the Yangtse), and is called *Ya-chio-tze*, "duck's foot," on account of the shape of the leaves. At the beginning of the Sung dynasty (1000 A.D.), the fruit was taken as tribute, and was then called *Yin-hsing*, "silver apricot," from its resemblance to a small apricot with a white kernel. In the *Chih-Wu-Ming*, xxxi. 27, there is a good figure of the foliage and fruit; and the statement is made that in order to obtain fruit the tree should be planted on the sides of ponds.

At present it occurs planted in the vicinity of temples in China, Japan, and Corea. It has always been the custom of the Chinese to preserve portions of the natural forest around their temples; and in this way many indigenous species have been preserved that otherwise would have perished with the spread of agriculture and the destruction of the forests for firewood and timber, in all districts traversed by waterways. Most of the curious conifers in China and Japan have a very limited distribution, and Ginkgo is probably no exception; though it is possible that it may still exist in the region indicated above.

I have never seen any remarkable specimens in China; but Bunge[1] says that he saw one at Peking, of prodigious height and 40 feet in circumference.

In Japan Elwes says that it is planted occasionally in temple courts, gardens, and parks. He did not see any very large specimen of the tree, the best being one in the court of the Nishi Hongagi temple at Kioto, which was of no great height, but had a bole about 15 feet in girth at 3 feet, where it divided into many wide-spreading branches which covered an area of 90 paces in circumference. This tree had green leaves and buds on the old wood of the trunk close to the ground, which he did not notice in other places.

Rein[2] says that the largest he knew of is at the temple of Kozenji near Tokyo, and this in 1884 was 7.55 metres in girth, and according to Lehman about 32 metres high. There is also one in the Shiba park, which in 1874 was 6.30 metres in girth. The tree is sometimes grown in a dwarf state in pots, but does not seem to be a favourite in Japan. The wood is somewhat like that of maple in grain, of a yellowish colour, fine grained, but not especially valued, though it is used for making chess boards and chessmen, chopping blocks, and as a groundwork for lacquer ware. The nuts are sometimes eaten boiled or roasted, but are not much thought of.

Ginkgo was first made known to Europeans by Kaempfer,[3] who discovered it in Japan in 1690, and published in 1712 a description with a good figure of the foliage and fruit. Pallas[4] visited the market town of Mai-mai-cheng, opposite Kiachta, in 1772, and saw there Ginkgo fruit for sale which had been brought from Peking.

---

[1] Bunge, in *Bull. Soc. d'Agric. du Depart. de l'Herault*, 1833.
[2] Rein, *Industries of Japan*. [3] Kaempfer, *Amœnitates Exoticæ*, 811.
[4] Pallas, *Reisen durch versch. Provinzen des Russischen Reiches*, 1768-1773, vol. iii.

Fortune[1] mentions that the tree grows to a very large size in the Shanghai district, and in the northern part of the Chekiang province. The Japanese name Ginkgo is their pronunciation of the Chinese *yin-kuo*, "silver fruit"; but the common name in Japan is *i-cho*.

INTRODUCTION

The tree was introduced into Europe about 1730, being first planted in the Botanic Garden at Utrecht. Jacquin brought it into the Botanic Garden at Vienna sometime after 1768. It was introduced into England about 1754; and into the Unites States in 1784, by W. Hamilton, who planted it in his garden at Woodlawn, near Philadelphia. It first flowered in Europe at Kew in 1795. Female flowers were first noticed by De Candolle in 1814 on a tree at Bourdigny near Geneva. Scions of this tree were grafted on a male tree in the Botanic Garden of Montpellier; and perfect fruit was produced by it for the first time in Europe in 1835.

CULTIVATION

Ginkgo is easily raised from seeds, which retain their vitality for some months. Female plants may be obtained by grafting. It is easily transplanted, even when of a large size. Trees of over 40 feet high have been successfully moved. It thrives in deep, well-drained, rich soil. It is useful for planting in towns, as it is free from the attacks of insects and fungi; and the hard leathery leaves resist the smoke of cities. It may also be freely pruned. It is of course best propagated by seed; but layers and cuttings may be employed in certain cases. Falconer (*loc. cit.*) says that it is not readily propagated by cuttings, and that it took two years to root a cutting in the gardens at Glen Core (U.S.A.). Pyramidal forms can be obtained by careful selection, and the broad-leaved variety by careful grafting. The Ginkgo is well adapted for cultivation in tubs or vases, and may then be trained either as a pyramid or a bush.

The tree has a formal appearance when young, and is not really beautiful till it attains a fair age. The peculiar form of the leaves renders it a striking object. The foliage, just before it falls in autumn, turns a bright yellow[2] colour, which makes it very effective in that season, but only for a few days, as the defoliation is very rapid.

REMARKABLE TREES

Ginkgo is perfectly hardy in England, and, as a lawn tree, is seen to great advantage. Many trees of considerable size occur in different parts of the country. The best known one is that at Kew, of which a photograph is given (Plate 21). In 1888 it was (measured by Mr. Nicholson) 56 feet in height, with a girth of 9 feet at

---

[1] See Fortune, *Wanderings in China*, 118, 251; *Residence among the Chinese*, 140, 348, 363; *Yedo and Peking*, 59.

[2] There is no trace of red in the autumnal tint, as is usual in other trees in their leaves before they fall. The tint in Ginkgo depends entirely on the yellow coloration of the disorganised chlorophyll corpuscles, and forms a beautiful object for the microscope.

a yard from the ground. It has a double stem, and in 1904 had increased to 62 feet high by 10 feet 4 inches in girth. Other remarkable trees near London[1] are :—

One at Chiswick House, which measured in 1889, 57 feet by 6½ feet, and in 1903, 62 feet by 6 feet 11 inches; and another at Cutbush's Nursery, Highgate, which was in 1903 56 feet high by 4½ feet in girth.

Ginkgo trees may be seen in the following places in London :—Victoria Park, Telegraph Hill, Lincoln's Inn Fields, Waterlow Park, Southwark Park.

At Grove Park, Herts, a tree measured in 1904 68 feet high by 8 feet 5 inches in girth.

At Bank House, Wisbech, the residence of Alexander Peckover, Esq., there is a tree which was 65 feet high and 7 feet in girth in 1904.

There is a very fine tree[2] at Frogmore, Windsor, which in 1904 measured 74 feet by 9 feet 3 inches, but divides into four stems (Plate 22).

At Barton, Suffolk, a tree planted in 1825 measured in 1904 50 feet by 2 feet 5 inches.

At Sherborne, Dorset, a tree 70 feet by 7 feet 7 inches in 1884.

At Melbury, Dorchester, the tallest tree in England is said to occur, being stated to be over 80 feet in height.[3] The tree at Panshanger[3] is reported to be 70 feet high by 10 feet at 1 foot above the ground. At Longleat[3] there is a tree 71 feet by 9½ feet girth at 1 foot above the soil.

At Cobham Park, Kent, a tree 68 feet by 9 feet 4 inches.

At Badminton, Gloucestershire, a pair of symmetrical trees each about 50 feet by 5 feet.

At Blaize Castle, near Bristol, there is a good tree, of which Lord Ducie has kindly sent a photograph and a letter from Miss Harford, dated December 1903, which states :—" The Salisburia is, I am glad to say, in perfect condition, and a very fine graceful tree. Its height, measured last summer, was 72 feet. I have always heard that the one at Kew (which is not nearly so well grown) and the one in the Bishop's garden at Wells came over from Japan in the same ship as our tree."[4]

In Wales the finest tree that we know of is at Margam Park, Glamorganshire, the residence of Miss Talbot, which in 1904 was about 70 feet high and 6 feet in girth.

We have not heard of any fine specimens in Scotland or Ireland.

A curious form of the Ginkgo tree is reported[5] to occur at Cookham Grove, Berkshire. This tree grows within 10 feet of the river wall, which surrounds the lawn, and when there is high water the roots are under water for several days at a time. The bole is only 2 feet in height, but measures 4½ feet in girth; at that point it breaks into many branches, some going upright to a distance of over 30 feet, while others grow almost horizontally, the spread of the branches being 45 feet.

---

[1] The well-known trees in the Chelsea Botanic Garden and in High Street, Brentford, are now mere wrecks.

[2] Figured in *Garden*, 1904, lxvi. 344.

[3] *Flora and Sylva*, ii. (1904), p. 357.

[4] Elwes has since seen and measured this tree, which he made to be 68 feet by 9 feet 3 inches, with a bole about 12 feet high.

[5] *Gard. Chron.* 1886, xxv. 53.

Much finer trees occur on the Continent than those in England; and it is evident that while the tree is healthy and hardy in this country, it requires hotter summers and colder winters to attain its best development and ripen fruit. A fine pair, male and female, stand in the old Botanic Garden of Geneva, where they were planted in 1815. They were measured by Elwes in 1905, when the male tree was 86 feet by 4 feet 10 inches, with a straight upright habit, the female, which bears good seed, was considerably smaller. A famous specimen in the garden adjoining the palace of the Grand Duke of Baden at Carlsruhe measured, in 1884, 84 feet, with a diameter of 25 inches at 3 feet from the ground. Beissner[1] says trees occur in this garden of $25\frac{1}{2}$ and 30 metres high, with stem diameters of 1.90 and 1.80 metres. The finest tree in Europe is probably one mentioned by Beissner,[1] which stands in the Botanic Garden at Milan, and measures 40 metres high and 1.20 metre in diameter. There is also a noble specimen in the gardens of the Villa Carlotta on Lake Como.   (A. H.)

---

[1] Beissner, *Nadelholzkunde*, 1891, pp. 191, 192.   One of the trees at Carlsruhe is figured in *Gartenwelt*, iv. 44, p. 520.

LIRIODENDRON

Liriodendron,[1] Linnæus, *Sp. Pl.* 535 (1753); Bentham et Hooker, *Gen. Pl.* i. 19 (1862).

TREES, several extinct and two living species, belonging to the Natural order Magnoliaceæ, with deciduous, alternate, stalked, saddle-shaped, or lyrate leaves. Flowers: solitary, terminal, stalked, regular, enclosed in bud in a 2-valved spathe, which falls off when the flower opens. Floral receptacle: cylindro-conic, bearing from below upwards 3 imbricated petaloid sepals, 6 petals imbricated in two rows, numerous stamens, with anthers dehiscing outwardly by longitudinal slits, and a spindle-shaped column of numerous densely imbricated independent carpels. Each carpel is a 1-celled ovary, containing 2 ovules, and terminating in a style with stigmatic papillæ at its apex. Fruit: a cone of samaræ, falling off the receptacle when ripe, each containing 1 or 2 seeds.

Liriodendron appeared in the Cretaceous epoch, and numerous fossil species have been found in North America and Europe in the Tertiary period. Of the two now living, one occurs in the eastern half of the United States and Canada, the other is a native of Central China.

---

[1] *Liriodendrum* is the spelling used by Linnæus in his earlier descriptions of the genus in *Corollarium Gen. Pl.* 9 (1737), and *Hort. Cliff.* 223 (1737); but the form given above is the one now always adopted.

LIRIODENDRON CHINENSE, CHINESE TULIP TREE

Liriodendron chinense, Sargent, *Trees and Shrubs*, iii. 103, Pl. lii. (1903) ; Hemsley in Hook. *Ic. Pl.* t.
   2785 (1905).
Liriodendron chinense, Hemsley, *Gard. Chron.* 1903, p. 370.
Liriodendron tulipifera, L., var.? *chinensis*, Hemsley, *Jour. Linn. Soc.* xxiii. 25 (1886).

The Chinese tulip tree was discovered by Shearer[1] and Maries[2] in the Lushan
mountains near Kiukiang, on the Yangtse, and was afterwards found by me growing
plentifully in the mountain woods both north and south of Ichang, in Hupeh, at
3000 to 6000 feet altitude.   Von Rosthorn[3] found it farther west, at Nan-ch'uan
in Szechuan.   It does not occur on the lower levels, and is essentially a tree of the
mountains bounding the valley of the Yangtse, from 107° to 116° E. longitude, and
from 29° to 32° N. latitude.   I never saw any large specimens, and it does not attain,
so far as is known, the size of the American species.   Von Rosthorn records it as
about 50 feet in height.   Maries notes it as a fine spreading tree occurring at a
temple near Kiukiang.   It was introduced in 1901 into cultivation from Hupeh by
Wilson, who collected for Messrs. Veitch ; and young trees may be seen in their
nursery at Coombe Wood, and also at Kew.   These seedlings in January 1905 were
at Kew about 15 inches in height, and have stood without injury the cold of the last
few winters ; but it is too soon yet to decide whether this species will turn out
to be hardy in this climate.

The Chinese tulip tree is almost indistinguishable in foliage from the American
tree, but as a rule the leaves are more glaucous on the under surface, and the
lobing is deeper and more obtuse.   The flowers are greenish in colour and smaller
in size than those of *Liriodendron tulipifera*.   Moreover, the narrow petals spread
out when fully open, and have not a tulip shape.   The carpels are consolidated, so
as to appear like a solid column, and are obtuse at the apex when ripe.   In the
American species the carpels are free from each other at an early stage, and have
when ripe acute recurved tips.

In winter there is little to distinguish the two species, except that in
Liriodendron chinense the twigs are grey (not shining brown), the buds come off at
a very acute angle, and the leaf scars are oboval and not truly circular as they are in
the common species.

The Chinese call the tree *Wo-ch'ang-ch'iu, i.e.* "goose-foot Catalpa," from the
shape of the leaves, but the tree is of no economic importance with them.   It
apparently regenerates readily from the stool, as I found it, where the wood-cutters
had been at work, as strong coppice shoots with enormous leaves, more than a foot
across.

[1] In 1875.   See L. M. Moore in *Jour. Bot.* 1875, p. 225.
[2] In 1878.   See Hemsley in *Gard. Chron.* 1889, vi. 718.   [3] Diels, *Flora von Central China*, 322 (1901).

LIRIODENDRON TULIPIFERA, Tulip Tree[1]

Liriodendron tulipifera, Linnæus, *Sp. Pl.* 1st ed. 535 (1753); Loudon, *Arb. et Frut. Brit.* i. 284 (1838); Sargent, *Silva of N. America*, i. 19, tt. 13, 14 (1891).

A lofty tree, attaining in America in the most favourable conditions a height of 190 feet, and a stem diameter of 10 feet. Bark grey and smooth in young trees, becoming darker in colour and furrowed in old trees. Roots, fleshy with pale brown bark, having an aromatic odour and pungent taste.

Leaves extremely variable in shape, but generally saddle-shaped or lyrate in outline, with a rounded or cordate base, and a truncate emarginate apex,[2] the midrib being prolonged into a short bristle. Sometimes they are quite entire, but are more usually lobed, the lobes varying from 2-6 or even 8 in number, and often ending in a point. Venation pinnate. The leaves are 3-5 inches in length and in breadth, dark green and smooth above, lighter in colour and minutely pubescent underneath. Stalks about as long as the blades, angled and slender, so that the leaves quiver with any movement of the air. In autumn they turn bright yellow in colour, and give the tree a handsome appearance.

Two lateral stipules[3] occur on the twigs, attached a little higher up than the insertion of the leaf-stalk. These are the scales which have formed the buds of the previous winter; and, as a rule, they shrivel up and fall off when the young leaves are fully matured; but some of them remain on vigorous shoots till late in summer.

The flowers resemble a tulip in shape, being $1\frac{1}{2}$-2 inches long, with a width of 2 inches at the summit. The petals are greenish white, with an orange-coloured band at the base, which secretes nectar attractive to bees. These visit the trees in myriads in May, the flowering season in Illinois.

The fruit, light brown in colour, is a cone made up of a large number (about 70) of ripe carpels, which consist of a 4-ribbed pericarp surmounted by a flattened woody wing (the enlarged style). The wing may carry the seed by currents of air 300 or 400 feet from the parent tree. The carpels remain on the tree till thoroughly dry, some usually persisting throughout the winter on the receptacle, a few falling at a time as the wind dislodges them. The outer ones are nearly always sterile. The carpels will float in water for nearly a year without sinking; and this may explain the distribution of the tree along the banks of rivers. The seed has a fleshy albumen, in the summit of which is situated a minute embryo.

Seedling. — The seedling has two aerial short-stalked oval cotyledons about

[1] Usually called "Yellow Poplar" in the United States, "White-wood" also being a name in use amongst the western lumbermen.

[2] Lubbock, in *Trans. Linn. Soc. (Bot.)* xxiv. 84, ascribes the form of the leaves to the way in which they are packed in the bud.

[3] Occasionally the stipules are attached as wings to the leaf-stalk either near its base or higher up; and in rare cases they even unite with the base of the leaf-blade, appearing then to be extra lower lobes of the leaf itself. For accounts of these peculiar stipules and remarkable forms of leaves occurring in tulip trees, see E. W. Bury in *Bull. Torrey Bot. Club*, 1901, p. 493, and in *Torreya*, 1901, p. 105, and 1902, p. 33.

$\frac{1}{2}$ inch in length. Above these on the stem follow the true leaves, the first and second orbicular in outline ; the third and fourth showing lobes ; all have long slender petioles. The first year's growth terminates in a bud just above the insertion of the fourth leaf. The primary root gives off a good many lateral fibres, which are delicate and brittle. Seedlings which germinated at Colesborne early in June were 3-4$\frac{1}{2}$ inches high in August, with roots of about the same length or slightly shorter. According to Elwes there was no marked tendency to form a tap-root in any of the specimens which he examined.

VARIETIES

Several forms are in cultivation, which differ from the wild tree in habit, in form and colour of the leaves, and in colour of the flowers.

1. Var. *pyramidalis*, Lavallée.—Tree with erect branches, forming a narrow pyramid, like the fastigiate oak.

2. Var. *integrifolia*, Kirch.—Leaves rounded at the base and without lobes. In this form, the shape of the leaves of seedling trees is preserved.

3. Var. *obtusiloba*, Pursh.—Leaves with only one rounded lobe on each side of the base.

4. Var. *heterophylla*.—Foliage variable ; some leaves being entire, others with lobes, which are acute or obtuse.

5. Var. *crispa*.—Leaves with undulate margins.

6. Var. *variegata*.[1]—Forms with variegated leaves, of which several sub-varieties have received names, as *argenteo-variegata, aureo-variegata, medio-picta*. That known as *aureo-marginata*[2] in which the edges of the leaves are yellow is the best.

7. Var. *aurea*.—Flowers yellow.

IDENTIFICATION

In summer, the shape of the leaves is unmistakable, resembling those of no other hardy tree : the variety *integrifolia*, though without lobes, preserves the truncate, slightly emarginate apex, in the centre of which may be seen the midrib prolonged as a short bristle.

In winter, the twigs and buds are very characteristic. Buds : terminal, larger than the lateral, which are alternate on the twigs, and arise from them at an angle of 45°. They are stalked, glaucous, glabrous, composed of 2 stipules joined together by their edges, forming a closed sac, in which is contained the young shoot ;[3] and on opening it a leaf will be seen embracing an interior bud. It is folded on its mid-rib with the stalk bent like a hook, bringing the apex of the leaf to the base of the bud. The twigs are glabrous, shining brown or slightly hoary, and marked by stipular rings just above the leaf-scars, which are circular, placed obliquely on prominent cushions, and dotted like a sieve with cicatrices of the fibrous bundles. The lenticels are few

---

[1] The variegated form in which the yellow marking occurs as irregular blotches in the central part of the leaves is well depicted in Lemaire, *Illust. Horticole*, xv. t. 571 (1868).

[2] A good figure of the variety is given in *Flore des Serres*, xix. 2025 (1873).

[3] Within the outer bud or sac are contained several younger buds, one within the other, each with a folded leaf.

and minute. The pith is solid, but not continuous, being interrupted by woody cross-partitions. (A. H.)

DISTRIBUTION

In Canada the tulip tree occurs[1] in rich soil in the western peninsula of Ontario, from Hamilton to Huron Co. It forms a noble tree in the thick forest west of St. Thomas, and has been found in Nova Scotia.[2]

In New England it occurs in the valley of the Hoosac river, Mass., in the Connecticut river valley, and in Rhode Island, where it is frequent.[3]

It extends west to Southern Michigan as far north as Grand river, southward through all the States east of the Mississippi to Alabama, attaining its maximum size in the valleys of the Ohio river and its tributaries, and in the foot-hills and valleys of the Southern Alleghany mountains, in Tennessee, Kentucky, and North Carolina. West of the Mississippi it occurs commonly, though not in the south-eastern parts of Missouri and Arkansas. Its southern limit appears to be in Northern Florida, Southern Alabama, and Mississippi.

Sargent says of this tree that it is one of the largest and most beautiful trees of the American forest, only surpassed in the Eastern States by the occidental plane and the deciduous cypress.

It sometimes attains in the deep river bottoms and warm, damp, summer climate of Southern Indiana a height of 160-190 feet, with a straight trunk 8-10 feet in diameter clear of branches for 80-100 feet from the ground. Individuals 100-150 feet tall with trunks 5-6 feet in diameter are still common. The branches, which are small and short in proportion to the trunk, give this tree a pyramidal habit, except in the case of old or very large individuals, on which the head is spreading.

I have seen it growing in the neighbourhood of Boston, where, however, it did not seem to attain as large a size as in the south of England, and where seedlings do not come up freely so far as I saw. Near the gate of the Arnold Arboretum the largest tree, about 70 years old, was 85 feet high by 8 feet 6 in girth.

In Druidhill Park, Baltimore, it becomes a much finer tree, and surpassed in height any other species growing there. The tallest I saw was in a shady dingle, and measured 125 feet by 11 feet, with a straight clean stem. Older trees had rough bark coming off in scales.

In the mountains of North Carolina, at Biltmore, I saw much larger trees, and to give an idea of its development in this region I figure (Plate 24) a tree from a photograph[4] kindly sent me by Mr. W. Ashe of the North Carolina Geological Survey, taken in the winter. This tree, which stood in Yancey Co., North Carolina, was a very characteristic specimen, more than 160 feet high and 6 feet in diameter at 5 feet from the ground. The smaller timber having been cut from around it only a few years previously, the form of the tree is perfectly typical, and shows the charac-

[1] Macoun, *Cat. Canadian Plants*, pt. i. 28 (1883).
[2] G. Lawson, *Proc. N.S. Inst. Science*, 86 (1891).
[3] Dame and Brookes, *Trees of New England*, 104 (1902).
[4] Pinchot and Ashe, *Bull. No. 6, North Carolina Geol. Surv.* pt. ii. (1898).

teristic sharp angles made by the smaller erect branches with the larger horizontal limbs.

Another photograph sent me by Mr. Ashe shows a group of Liriodendron, in the forests of Transylvania Co., N.C., 120-140 feet high and 4-5 feet in diameter, associated with *Quercus rubra* and *Betula lutea* which are not so tall. This magnificent forest is, like most of those accessible to the lumbermen, rapidly decreasing in area and beauty, owing to the growing demand for timber.

For further details of the distribution in North Carolina refer to Pinchot and Ashe's admirable account, pp. 39-41, and to a paper by Overton Price on " Practical Forestry in the Southern Appalachians."[1]

The largest trees of this species, however, have been recorded by Professor R. Ridgway[2] from Southern Indiana and Illinois, near Mount Carmel, Illinois, which I had the pleasure of visiting under the guidance of Dr. J. Schneck in September 1904. Though the largest trees recorded by him have now been cut, reliable measurements were taken of a tulip tree which reached the astonishing height of 190 feet, exceeding that of any non-coniferous tree recorded in the temperate regions of the northern hemisphere. Another tree cut " 8 miles east of Vincennes, was 8 feet across the top of the stump, which was solid to the centre ; the last cut was 63 feet from the first, and the trunk made 80,000 shingles." The soil here is an exceedingly rich, deep alluvium, and the climate in summer very hot and moist.

It is stated in *Garden and Forest*, 1897, p. 458, that at the Nashville Exhibition a log of this tree was shown by the Nashville, Chattanooga, and St. Louis Railroad Company, which measured 42 feet long, 10 feet 4 inches in diameter at the butt, and 7 feet at the smaller end, containing 1260 cubic feet of timber, and about 600 years old.

INTRODUCTION

The tulip tree was probably introduced, according to Evelyn,[3] by John Tradescant about the middle of the seventeenth century, but this is somewhat uncertain, though it was grown by Bishop Compton at Fulham in 1688.

According to Hunter the tree which first flowered in England was in the gardens of the Earl of Peterborough at Parsons Green, Fulham, and this he describes in 1776 as "an old tree quite destroyed by others which overhang it." At that time there were also some trees of great bulk at Wilton, the seat of the Earl of Pembroke in Wilts.

CULTIVATION

Though the tree can be propagated by means of layers, and in the case of varieties by grafting, yet as seeds are easily procured from the United States it is much better to raise it from seed. Cobbett, who was a great admirer of the tulip tree, gives a long account of it, and of the best means of raising it,[4] and says that if sown in May, which he thinks the best time, it will germinate in the following May,

---

[1] *Yearbook U.S. Dept. of Agric.* (1900).
[2] *Notes on Trees of Lower Wabash*, Proc. U.S. Nat. Hist. Mus. 1882, p. 49 ; 1894, p. 411.
[3] Evelyn's *Silva*, 214. Ed. Hunter (1776).   [4] *Woodlands*, par. 523 (1823).

but that if sown in autumn, part will come up in the next spring and part in the following year.

Dawson in an excellent paper on the Propagation of Trees from Seed,[1] says, "The tulip tree invariably takes two years, and as the proportion of good seed is as 1 to 10, it should be sown very thickly to ensure even an ordinary crop."

Probably this opinion was based on his experience with seeds grown in New England, where they do not ripen so well as they do in the south, for my own experience, gained by sowing seeds received from Meehan of Philadelphia, is different. In the spring of 1903 I sowed part of the seeds in a greenhouse, where they began to germinate six weeks later. Of those sown in the open ground, perhaps 10 per cent germinated in June. The following summer was cold and wet, and the seedlings in the open ground made slow progress, being only 2–3 inches high in the autumn, whilst those kept under glass were from 6-15 inches high at the same time. The young wood seems to ripen better than that of most North American trees and, as the spring of 1904 was favourable, they were not checked by frost. But the seedlings are difficult to transplant, owing to the fleshy and brittle nature of their roots, and are therefore best kept in a box or large pot till they are two years old, when the roots should be trimmed and planted out in deep sandy soil, and watered the first year; after this they should be transplanted frequently until large enough to put in their permanent situation, and if tall and straight grown trees are desired the young trees must be very carefully pruned, as like the Magnolia they do not thrive so well if large branches are cut off.

The tulip tree rarely ripens its seed in England, and that which I got from a tree at Westonbirt in Gloucestershire in 1901 did not germinate. But I am informed by Mr. A. C. Forbes, that a self-sown tulip tree is growing in the sand walk at Longleat, and Colonel Thynne confirms this in December 1904, when he tells me the young tree is 8 feet high. This, however, is the only instance I know of in England where natural reproduction has occurred.

SOIL AND SITUATION

The tree requires a deep, moist, rich soil to bring it to perfection, preferring heavy land to light, and apparently disliking lime in the soil. It probably prefers a moderate amount of shade when young, and would be more likely to grow tall and straight if surrounded by other trees. But isolated trees sometimes grow with a clean straight stem, as at Leonardslee in Sussex (see below) even on dry soil.

In the *Gardeners' Chronicle* for 1879 there was much correspondence on the merits of this tree for general cultivation in England, from which I extract the following particulars, which will be valuable to intending planters.

Most of the correspondents agree that it grows best on heavy soil, inclining to clay, or with a clay subsoil. Sir W. Thiselton Dyer says it does not do well on the light, dry soil of Kew Gardens.

[1] *Trans. Mass. Hort. Soc.* 1885, p. 152.

Mr. Bullen says that it grows well in heavy clay in the damp and smoky climate of Glasgow, and a tree is mentioned at the Grove, Stanmore, on damp, gravelly clay, which in 1879 was 77 feet high by 9½ in girth.

The tulip tree has been much recommended for planting in towns, and specimens may be seen in London at Victoria Park, Manor House Gardens, Lincoln's Inn Fields, Waterloo Park, Clissold Park, etc.

Mr. Hovey says that in America it is not so much planted for ornament as it deserves to be, presumably because American planters desire a quick effect, and that it does not transplant well after it is 4-6 feet high ; but that it grows on gravel, sand, peat, or clay, and is not very particular in that climate as to soil.   He has known it grow 30 feet high and more in 20 years.

It is very liable to be attacked by rabbits, which eat the bark even of large trees, and I have seen several which have been killed or much injured in this way.

REMARKABLE TREES

Though this tree is one of the handsomest when in flower, stateliest in habit, and most beautiful in the autumn tints of its leaves, it is not now planted in England nearly as much as it was a hundred years and more ago, having, like so many other fine hard-wooded trees, been supplanted by conifers and flowering shrubs, which are easier to raise and more profitable to the nurserymen, who now appear to cater rather for the requirements of owners of villas and small gardens than for those of larger places.   But though the tulip tree loves a hot summer, it endures the most severe winter frosts of our climate without injury, and in a suitable soil grows in some parts of the southern counties, after it is once established, to a great size.

The largest living specimen I know of in England is at Woolbeding, in Sussex, the seat of Colonel Lascelles, and measures 105 feet by 17.   Though not so perfect in shape as some others, it is a very beautiful tree, and seemed, when I saw it in 1903, to be in good health.   It grows on a deep, alluvial, sandy soil, which suits plane trees and rhododendrons very well (Plate 25).

There was even a larger one at Stowe near Buckingham, which when I saw it in 1905 was dead, apparently barked at the base by rabbits.   It was at least 107 feet high, with a bole of about 30 feet, and a girth of 13 feet at 5 feet, and 21 feet 4 inches at the ground.

Another very fine tree is at Leonardslee, near Horsham, the seat of Sir Edmund Loder, Bart., also in Sussex, and is growing at an elevation of 400-500 feet on soil which, though very favourable to rhododendrons, is too poor to grow either oak, birch, or larch to the same size in the same time.   Sir E. Loder tells me that the tree cannot be more than 90 years old, and it is now 97 feet high, with a perfectly clean, straight trunk 25-30 feet high, which towers above all the native trees of the district (Plate 27).

At Horsham Park, the residence of R. H. Hurst, Esq., is a very fine and symmetrical tree which I measured rather hastily, as over 100 feet in height by 15 in girth.

Another very remarkable tree (Plate 26) is the one at Killerton, in Devonshire, which I am sorry to hear has suffered severely in the gale of September 1903. This tree must be one of the oldest now living, as Sir C. T. D. Acland tells me that in a picture of his house, taken early in the last century, it seems nearly as tall as at present, and it is mentioned by Loudon as being 63 feet high in 1843. When I measured it in 1902 it was 80 by 15 feet, with a bole about 18 feet long, and must have contained nearly 300 feet of timber.

A very fine tulip tree, on heavier, damper soil at Strathfieldsaye, Berkshire, the seat of the Duke of Wellington, measures 105 feet by 12 ; and though not such a well-shaped tree as the one at Leonardslee is of the same type.

The tree which Loudon refers to as being the tallest known to him at Syon, was, in 1844, 76 feet high, at about 76 years of age, but this is now dead, as is the old tree at Fulham Palace mentioned by Loudon, which he estimated at 150 years of age.

At Bury House, Lower Edmonton, there is a magnificent tree which John W. Ford, Esq., informs us is thoroughly sound and in perfect health. He estimates it to be 70 to 75 feet in height, the girth 5 feet from the ground being 17 feet 4 inches. The bole at 13 feet divides into five limbs, of which the biggest are 5 feet round. The soil is splendid, being brick earth.

At Deepdene, Dorking, there is a fine tree on the lawn, which in February 1904 was 83 feet high by 14 feet in girth.

At Petworth, the seat of Lord Leconfield, there is a curious old tree which has an immense burry trunk 17 feet in girth.

A tree was recorded at Longleat in 1877 as being 106 feet high and 10 feet in girth, but this, as I learn from the Marquess of Bath, is now dead, though one or two other large specimens remain.

There is a very fine tree at Margam, in Pembrokeshire, which, as measured in 1904, is 92 feet high by 13 feet 6 inches at 6 feet from the ground, with a spread of branches 57 feet in diameter.

An immense tree at Esher Place, Surrey, is mentioned by Mr. Goldring as having a girth of 22 feet.

At Barton, Suffolk, two trees[1] were planted in 1832. They first flowered in 1843. In the year following the severe winter of 1860 no flowers were produced, but the foliage was as good as usual. In 1904 these two trees had both attained the same height—79 feet ; one having a girth of 7 feet 2 inches at 5 feet above the ground ; the other divided into two stems at a point 2 feet from the ground where the girth was 10 feet 4 inches. The soil at Barton is good, consisting of 2 or 3 feet of loam resting on boulder clay.

At Ashby St. Ledgers, Rugby, the seat of the Hon. Ivor Guest, there is a good tree[2] which measured 80 feet in height by 16½ feet in girth in 1900. This tree breaks into three stems at a little above 4 feet from the ground, and the girth is taken below this point.

At Hampton Court, Herefordshire, a tree[3] on the lawn in 1879 was 80 feet

[1] Bunbury, *Arboretum Notes*, 60.　　　　[2] Letter to Curator at Kew.

[3] *Garden*, 1890, xxxviii. 178. The measurements refer to 1879, according to a note in *Woods and Forests*, April 23, 1884.

high by 12 feet 7 inches in girth, with an estimated cubic contents of timber of 223 feet. When I measured it in 1905 it was 95 by 13 feet, but the top and trunk were decaying.

At Erlestoke Park, Wiltshire, a tree,[1] growing near the bank of a lake, was 80 feet high by 14 feet in girth at 4 feet from the ground in 1902.

The following records from Hampshire were reported in *Woods and Forests* :[2]— North Stoneham Churchyard, near Southampton, a tree 12 feet 10 inches in girth ; Cranbury House, near Winchester, a tree 11 feet 9 inches in girth ; at Gramwell's Meadow, east of East Tytherley Manorhouse, near Romsey, a tree 85 feet high by 10 feet 5 inches in girth, with a stem free from burrs, planted in 1780. These measurements were taken in 1884. At Hale Park, in 1879, there was a tree 75 feet high with a short bole of 4 feet, girthing 18 feet 3 inches.

The finest tree at Kew, 70 feet high in 1844, is gone, but there still exists a well-proportioned specimen[3] which stands at the end of the rhododendron dell. It is now (1905) 79 feet high by 9 feet 9 inches in girth. It produces fruit freely every year, but the seeds are always poorly developed and infertile.

In Scotland a tree was mentioned by Loudon as growing at The Hirsel, Coldstream, the seat of the Earl of Home, which was at that time 100 years old and 20 feet in girth 3 feet from the ground. I was informed by Mr. Cairns, head gardener at the Hirsel, that in 1903 it was slowly decaying, some of the larger branches being gone, but that what remained carry a large amount of healthy foliage, and flowers more or less every year.

At Drummonie Castle, Perthshire, formerly a seat of the Lords Oliphant, Hunter[4] mentions a tree 8 feet in girth at 5 feet, and another at Gorthy Castle,[5] girthing 9 feet 7 inches at 3 feet, which had been a good deal injured by cattle grazing in the park. He also (p. 400) speaks of a large tree at Castle Menzies, 10 feet in girth, but I did not see it on either of my visits to this interesting old place.

The tulip tree is not mentioned in the *Old and Remarkable Trees of Scotland*, but it grows at Gordon Castle, and even as far north as Dunrobin Castle in Sutherlandshire.

In the south-west of Scotland there do not appear to be any large trees, the biggest mentioned by Messrs. Renwick and M'Kay[6] being one at Auchendrane House, Ayrshire, which was, in September 1902, 53 feet by 5 feet 8 inches, and one at Doonside, Ayrshire, which was 46 feet 9 inches by 8 feet 1 inch.

At Jardine Hall, Lockerbie, a tree[7] measured in 1900 60 feet in height by 9 feet in girth.

At St. Mary's Isle, Kirkcudbright, a tulip tree[8] was, in 1892, 10 feet 9 inches in girth.

[1] *Gard. Chron.* 1902, xxxii. 61.　　[2] Issues of April 16 and 23, 1884.

[3] Figured in *Gard. Chron.* 1890, viii. 219, where it is stated in the text that the tulip tree bears pruning well, and that there is an avenue of clipped trees in one of the courts at Chatsworth.

[4] Hunter, *Woods, Forests, and Estates of Perthshire*, 145 (1883). This is apparently the tree mentioned in *Gard. Chron.* 1890, viii. 388, as being 60 feet in height then, and having recently flowered.

[5] *L.c.* p. 371.　　　　[6] Renwick and M'Kay, *Brit. Assoc. Handbk.* 131 (1901).

[7] *Garden*, 1890, xxxviii. 178.　　　　[8] M'Kay and Renwick, *Trans. Nat. Hist. Soc. Glasg.* Sept. 4, 1894, p. 17.

In Ireland large tulip trees are rare. There are two good specimens at Fota, which measured in 1903, one 87 feet high by 11 feet 7 inches in girth, the other 57 feet by 14 feet 7 inches.

In France the tulip tree, favoured by warmer summers, seems to thrive better, and attains a larger size than in England. Mouillefert[1] speaks of a tree at the Chateau de Frêne, near Chaulnes, in the department of Somme, which in 1899 was 38 metres in height by 5 in circumference. He also mentions having seen in 1902 at the Chateau de Cheverny, near Blois, tulip trees planted along the banks of a canal, which at 50-60 years of age measured 31 metres in height and 2 metres in girth at 5 feet from the ground, whilst plane trees of the same age close to them were only 24 metres high and 1.65 in girth.

He considers that in a suitable soil and situation such as the valleys in a granitic mountain range, or on damp, rich soils, in fact in such places as the ash, the poplar, and the plane thrive, this tree might be grown as a forest tree to produce valuable timber, or as copse wood, cut at 18 or 20 years of growth for turnery purposes.

Considering, however, the cost of raising this tree in the nursery, and its liability to suffer from autumn frost in a young state, I do not think the tree can be considered likely to become a forest tree in England, except possibly in a few choice situations in the south and south-west.

TIMBER

The timber of the tulip tree is now very much used in North America for many purposes, and is also largely imported to England under the name of white-wood, canary-wood, and yellow poplar. Stevenson says of it,[2] "Though classed among the light woods it is much heavier than that of the common poplar, its grain is equally fine but more compact, and the wood is easily wrought and polished. It is found strong and stiff enough for uses that require great solidity. The heart-wood, when separated from the sap and perfectly seasoned, long resists the influence of the air, and is said to be rarely attacked by insects. It is imported in the form of waney logs and in sawn planks of very fine dimensions, in which state it commands a price fully equal to that of the first quality of Quebec yellow pine.

Hough[3] speaks of it as "light, rather strong, with close straight grain, compact, easily worked, and yielding a satiny finish. Sap-wood nearly white, heart-wood of a light lemon-yellow colour, or sometimes of a light brownish tint—whence its two seemingly contradictory names, white and yellow poplar, the former referring to the sap-wood, the latter to the heart."

Sargent says it is light and soft, brittle and not strong, is readily worked, and does not easily split or shrink. The heart-wood is light yellow or brown, weighing when absolutely dry 26-36 lbs. to the cubic foot. Large canoes were formerly made from it by the Indians, and it is now extensively used in construction, for the

[1] *Traité de Sylviculture*, 467-468 (1903). [2] *The Trees of Commerce*, 96-103 (1902).
[3] *The American Woods*, pt. i. p. 40, t. 2 (1893).

interior finish of houses, and in boat-building, as well as for shingles, pumps, and wooden ware.

The only timber I know which it resembles closely in colour, texture, and grain, is that of *Magnolia acuminata*.[1]

Neither Stevenson, Hough, nor Stone, however, speak of a form of this timber known as "blistered poplar," which is occasionally found, as I believe, only in old trees, and which is sometimes imported in small quantities to Europe. This seems akin to the figured maple wood known as bird's eye maple, but has the figure in oblong patches from 2 inches long downwards, of a dark olive colour on a paler olive-green ground, and is one of the most ornamental woods I know, fit to be used in the finest cabinet work. I saw large planks of this variety in the Exhibition at St. Louis, and have had some of it worked into the panels of a screen.

The wood of the tulip tree grown in England seems to be nearly as good in quality as the imported timber, though not quite so pure in colour. From a tree which was cut at Highclere a plank was sent me by the kindness of the Earl of Carnarvon, which has been used in the same screen, and I have a large book-case of which the back is made of the imported wood, selected by an experienced cabinet-maker as best for the purpose.

Mouillefert says that in Paris its use is increasing for all purposes for which the wood of the lime and poplar is suitable, and that it has when fresh cut a pleasant smell of orange, which, however, is soon lost as it dries.   (H. J. E.)

---

[1] Mr. Weale tells me that the timber of this Magnolia, as well as that of *M. grandiflora* and *M. glauca*, come into the Liverpool market mixed with that of the tulip tree, and that though the two former may easily be distinguished by a person who knows them well, yet that *M. glauca* can only be identified with a lens, and that in consequence of this mixture, opinions differ as to the suitability of the wood for laying veneers upon. He thinks that if bone dry, the wood of the tulip tree is fit for this purpose, but not equal to that of American chestnut, American cherry, or Honduras mahogany, of which the latter is best. He also says that for pattern making Quebec yellow pine is distinctly superior, and worth from 1s. to 2s. a foot more.

PICEA

SPRUCE-FIRS

Picea, Link, *Abhandl. Akad. Wiss. Berlin*, 1827, 179 (1830); Bentham et Hooker, *Gen. Pl.* iii. 439 (1880); Masters, *Jour. Linn. Soc. (Bot.)* xxx. 28 (1893).

Abies, Linnæus, · *Gen. Pl.* 294 (in part) (1737); D. Don in Lambert, *Pinus*, vol. iii. (1837), ex Loudon, *Arb. et Frut.* iv. 2293 (1838).

THIS genus includes the spruce-firs, which in England, following the practice of Don and Loudon, are still often called Abies. However, all botanists in England, on the Continent, and in America apply the term Picea to the spruces, and Abies to the silver firs.

Tall evergreen trees belonging to the tribe Abietineæ of the order Coniferæ, with shoots of only one kind, bearing in spiral order peg-like projections ("pulvini"), from which the leaves arise singly. The needle-like leaves are either tetragonal or flattened in section, and persist for many years, rendering the foliage very dense. At the ends of the leading shoots there is a terminal bud, with 2-5 side buds directly under it; the buds are dry and not resinous.

Flowers monœcious. Male flowers solitary in the axils of the uppermost leaves, ovoid or cylindric, short-stalked, surrounded at the base with scale-like bracts, composed of numerous stamens spirally arranged, each with 2 pollen-sacs opening longitudinally, and a connective prolonged into a toothed crest. Pollen grains with 2 air-sacs. Female flowers solitary, terminal, erect, stalked, with a few empty scales at the base; composed of 2 series of scales, the bracts small and membranous, and the ovular scales bearing at their base 2 inverted ovules. Cones: generally becoming pendulous, but in certain species remaining erect or spreading; cylindrical or ovoid, with the bracts minute and concealed, and the scales enlarged and firm in texture, with entire or denticulate margins, and bearing on their inner surface 2 winged seeds. The cones are ripe in the first season, and after dispersal of the seed (the scales persisting on the axis) fall off in the following winter, or remain in some species much longer on the tree. The cotyledons are 5-15 in number, 3-sided, and serrate in margin.

Species of spruce occur in Europe, Asia Minor, the Caucasus, Siberia, Mongolia, China, Japan, the Himalayas, and in North America. The genus is marked out into two natural sections by the character of the leaves. These are defined by Willkomm as follows :—

1. *Eu-picea.*—True spruces.   Needles 4-sided and 4-angled, with stomata on all their surfaces.   The ripe cones are always pendulous.

2. *Omorica.*—Flat-leaved spruces.   Leaves 2-sided, flattened from above downwards, stomata being only borne on their dorsal surface.   Ripe cones pendent, horizontal, or erect.

Other divisions have been made, such as that of Link into two sections, *Genuinæ* and *Dehiscentes*; and that of Mayr into three sections, *Omorica* (not identical with Willkomm's section of the same name), *Morinda*, and *Casicta*; but it is most convenient to adopt Willkomm's divisions.

The arrangement of the leaves on *lateral branchlets* is different in the two sections.   All spruces agree in the disposition of the leaves arising from the upper side of such branchlets, as these always point forwards and cover the shoot.   But, in ordinary species, the leaves underneath, while they part into two lateral groups, alter little their direction, which is more or less forwards; and the under part of the stem is laid only partially bare.   In almost all the flat-leaved spruces, the leaves below part into two sets, which are directed outwards at right angles to the shoot, which is laid quite bare.   This arrangement differs from that of the yew and most silver firs, where the leaves are divided into two sets both above and below; and this distinction depends on the fact that in these spruces the stomata are on the dorsal surface of the leaf, whereas in the yew, etc., they are on the ventral surface; and in the effort to direct the stomata away from the light, a different arrangement results in the two cases.

The arrangement of the leaves on leader or upright branchlets is the same in all species of spruce, being radial, the leaves pointing outwards and slightly forwards. In certain species, as *P. Breweriana*, *P. Morinda*, the lateral branchlets are pendulous and not horizontal; and the leaves then are similarly arranged in both the lateral and the leader shoots.

The section *Eu-picea* will be dealt with in a later part.

KEY TO SECTION OMORICA.—The flat-leaved spruces are distinguished from the silver firs by the peg-like projections on the shoots, and from ordinary spruces by the flattened leaves with stomata only on their dorsal surface.

I. *Young shoots glabrous, yellow.*

 1. **Picea hondoensis.**   Central Japan.
  Buds broadly conical, with scales rounded in the margin, opening red.   Shoots of second year red.   Leaves thin, slightly keeled on both surfaces, blunt or ending in a short point.

 2. **Picea ajanensis.**   Manchuria, Amurland, Saghalien, Yezo.
  As in 1, but the buds open green, and shoots of the second year are yellow.

 3. **Picea sitchensis.**   Western North America.
  Buds ovoid with ovate obtuse scales.   Leaves deeply keeled on ventral green surface, almost convex on dorsal white surface, ending in very sharp, cartilaginous points.

4. **Picea morindoides.**[1]   Native country unknown.

Buds and scales ovate - obtuse.   Leaves linear, straight, slender, acuminate, terminating in a callous sharp tip, somewhat flattened and distinctly keeled on both sides, marked with two white lines on the upper surface, and dark bluish green on the under surface.   Leaves radially spreading on the branchlets.

II. *Young shoots pubescent with short hairs.*

5. **Picea Omorika.**   Servia and Bosnia.

Pubescence brown.   Buds ovate, conical, with outer scales ending in long subulate points.   Leaves flattened but thick, obtuse or ending in a short point.

6. **Picea Breweriana.**   Oregon, California.

Pubescence grey.   Buds ovoid, with outer scales ending in long points.   Leaves scarcely flattened, but convex above and below, keeled on dorsal surface, with midrib prominent on ventral green surface, and ending in a short point.   The leaves spread out in all directions on the shoot.

[1] A new species described by Rehder in Sargent, *Trees and Shrubs*, 95, t. 48 (1903).   It is only known as a tree growing in the arboretum of G. Allard at Angers.   I have seen no specimens and take the characters given above from Rehder. In habit it resembles *Picea Morinda*, the branches being pendulous.   The cones resemble those of *Picea Alcockiana.*

PICEA OMORIKA, Servian Spruce

Picea Omorika, Bolle, *Monatschrift des Vereines zur Beforderung des Gartenbaues*, 124 (1877); Masters, *Gard. Chron.* 1884, xxi. 308, 309, Figs. 56, 57, 58, and 1897, xxi. 153, Fig. 44; *Jour. Linn. Soc. (Bot.)* xxii. 203 (1886); Willkomm, *Forstliche Flora*, 99 (1897); Kent, in *Veitch's Man. Coniferæ*, 442 (1900); Richardson, *Edin. Bot. Garden, Notes*, No. 1 (1900); G. von Beck, *Die Vegetationsverhältnisse der Illyrischen Länder*, 286, 360, 440, 474 (1901).

Pinus Omorika, Pančic, *Eine Neue Conifere in den Oestlichen Alpen*, 4 (Belgrade, 1876); Masters, *Gard. Chron.* 1877, vii. 470, 620.

A tree with a tall, slender stem, said to attain 130 feet in height, with a girth of stem of only 4 feet, with short branches, forming a narrow pyramidal crown. The topmost branches are directed upwards, the middle ones are horizontally spreading, and the lower ones are pendulous, with their tips arching upwards. Bark brownish red, and scaling off in plates, the fragments often being heaped in quantity round the base of the tree. The leaves on vertical shoots stand out on all sides, but on horizontal shoots they point forwards on the upper side, being pseudo-distichous in three or four ranks on the lower side. They are flattened, 4-angled, straight, or curved to one side, $\frac{3}{4}$-1 inch long, linear, acute or obtuse with an apiculus, convex, and shining green on the ventral surface, marked with stomatic lines on each side of the prominent midrib of the dorsal surface.[1] They persist for 4 or 5 years.

The buds, ovoid-conic with brown, membranous scales, the outermost of which end in long subulate points, are produced chiefly near the end of the shoot; and in unfolding, the uppermost scales are pushed off as a cap. The dark brown hairs, which are conspicuous on the young shoots, persist on the older branchlets of even 3 or 4 years' growth in wild specimens.

The male flowers, which are partly solitary and partly whorled, are stalked, ovoid-cylindric, bright red, $\frac{1}{2}$-$\frac{3}{4}$ inch long, and are surrounded at the base by numerous membranous bracts.

Cones, shortly-stalked 2-2$\frac{1}{2}$ inches long, bluish black when young, dark-brown when ripe, clustered, the upper ones being directed upwards, while the middle ones are horizontal, and the lower ones pendulous. Scales almost orbicular in outline, broad and convex, streaked on the outer surface, with the margin slightly bent inwards, undulate and denticulate. Bract obovate-cuneate, minute. Seeds small, $\frac{1}{10}$-$\frac{1}{8}$ inch long, obovate, blackish brown, with a wing $\frac{1}{3}$ inch long, obovate in outline.

---

[1] On horizontal shoots, the leaves, by twisting movements on their bases, are inverted, so that the green surface is turned upwards and the stomatic surface downwards.

DISTRIBUTION

The Servian spruce was first made known to science by Pančić, who discovered it in south-western Servia, near the village of Zaovina, on 1st August 1875. Its area is a small one, occupying about 20 kilometres long by 15 kilometres wide on both sides of the Drina valley, the boundary between Servia and Bosnia. Here it occurs on limestone rocks at altitudes varying from 2700 to 5300 feet. It grows in small groves in the wetter places in the ravines, but it does not there reach such a height as it attains in the rockier parts of the mountains, where it forms part of the mixed forest of Austrian and common pines, common spruce, beech, and sycamore. Pure woods of Omorika occur at higher elevations, between 4700 and 5300 feet, where sub-alpine plants accompany it. Wettstein gives the following as the composition of the characteristic Omorika woods :—

Dominant Trees.—*Picea Omorika, Pinus sylvestris, Carpinus duinensis, Picea excelsa, Fagus sylvatica, Populus tremula, Abies pectinata, Ostrya carpinifolia, Salix sp., Pinus austriaca.*

Underwood. — *Corylus avellana, Cotinus coggygria, Spiræa cana,* with *Rhamnus fallax* and *Lonicera alpigena* at high altitudes.

Ground-herbage.—*Aspidium Filix-mas, lobatum,* and *angulare.*

Wettstein[1] says than an Omorika forest has a peculiar and gloomy aspect, the slender stems with their short branches and columnar or spindle-shaped crowns looking quite different from any other type of European forest. In mixed forests, the straight single stems, arising out of the general mass of the other trees, are equally peculiar.

Omorika seedlings and young trees are only found in exposed rocky situations, and in the bottoms of wet shaded ravines. The tree in the wild state is strictly confined to limestone soil, and never grows on the slate formation which is found in parts of the Drina valley, yet when cultivated, it does very well, at least in youth, on soils which are not calcareous.

The largest tree[2] recorded is one felled by Pančić, which measured 42.2 metres in height, and 0.385 metres in diameter. It showed 137 rings, and the width of the rings gradually diminished from 0.28 cm. in the 3rd decade to 0.04 cm. in the 14th decade. Pančić says that the tree has an inclination to grow with a spiral stem, and that it loses its branches up to about half its height, the largest of the branches being only about 2 metres in length. The cones are borne, according to him, upright on the topmost branches only, but elsewhere they hang down with their tips directed slightly upwards.

Pančić, in his first account of the tree, reports that he had heard on good authority of its occurrence in the mountains of Montenegro; it has since been

[1] *Sitzungsber. kais. Akad. d. Wiss.,* xcix. 503 ; *Oesterr. Bot. Zeitschr.* 1890, p. 357.
[2] Letter of Pančić, quoted in Stein's article on " Omorika " in *Gartenflora,* 1887, p. 14.

reported to occur also at Bellova in the Rhodope mountains in Bulgaria; but, so far as we can discover, these statements have not been confirmed. A fossil species which has been identified with the existing tree by Webber has been found in the interglacial deposits at Höttingen near Innsbruck in the Tyrol. An allied species, *Picea omorikoides*, Webber,[1] has been found at Aue in Saxony in a preglacial deposit which is of the same age as the Cromer forest bed on the coast of Norfolk. Lokowitz has also found near Mulhouse in Alsace some remains of a spruce in the middle Oligocene beds which resembles *Picea Omorika*.

In the herbarium at Kew there are specimens collected by V. Crucic on the Drina, and others with good cones gathered by Elwes at 2000 to 3000 feet altitude.

(A. H.)

I visited the valley of the Drina in Bosnia in 1900 on purpose to see this tree, and after driving a long day east from Serajevo, reached Rogatica, from where Herr Gschwind, the obliging forest officer of the district, was good enough to accompany me to Han Semec, a Gendarmerie station on the road to Visegrad, about 15 miles from Rogatica. Han Semec is at an elevation of 3800 feet, and is surrounded by beautiful forests of Austrian and Scots pines, spruce, silver fir, and beech.

The climate of the district is very cold in winter and warm in summer. The minimum temperature being $-33°$ Reaumur on 23rd December, $+30°$, the maximum on 7th July 1897, the snow lying as long as 4-5 months.[2] The rainfall in summer is heavy, amounting to 116.2 centimetres, which fell on 124 days, and the weather was wet most of the time I was there.

After passing through some beautiful mountain meadows and primæval forest of large spruce and silver fir mixed in places with beech and aspen, as well as small oaks and large birch, we came to the edge of a deep rocky ravine running down to the Drina valley. On the steep limestone cliffs overhanging this ravine, which are a favourite haunt of chamois, *Picea Omorika* was growing in clumps, and isolated trees occurred among common spruce, Scots and Austrian pine.

The branches are short and drooping as compared with those of common spruce, and the cones being found only near the top of the tree, we had to cut one down in order to procure fruiting specimens; on this I found young cones of the year, cones of last year which had not yet opened, and which, according to the forester, contained good seed only when there was turpentine exuding from them, and old cones which hang two or three years on the tree after shedding their seed. In habit and appearance the tree resembles the American *Picea alba* more than any tree I know, though its nearest botanical affinities are with *P. sitchensis* and *P. ajanensis*. Plate 28, which is from two of several photographs kindly sent me by Herr Othmar Reiser of the Landes-museum, Serajevo, Bosnia, gives an excellent idea of the forest and of individual trees.

The average size of the full-grown trees on these steep cliffs was not above 50-60 feet, with about 1 foot of diameter, but I found some measuring 80-90 feet high and 18 inches diameter. Young seedlings were scarce and difficult to find on the mossy rocks; but we collected 20 or 30 plants, of

---

[1] *Engler's Bot. Jahrb.* xxiv. 1898, Heft 4, 510, 504.
[2] Cf. *Met. Beob. Land Stationen in Bosnien* (1899).

which I brought the smallest home in a tin box alive, and planted the larger ones in the forester's garden at Han Semec. Those which I brought home have established themselves slowly but a quantity of seed received in the autumn germinated well in boxes, and in November 1905 was much larger than common spruce of the same age. They were quite uninjured by the severe frost of May 21, 1905, which injured the common spruce very severely, and on my limy soil are growing faster and more vigorously than any other species of Picea.

The tree appears to have been first distributed by Messrs. Fröbel of Zurich about 1884, and has been found quite hardy in England, as might be expected from the climate of its native country.

The finest specimen I know of in England is in the garden of W. H. Griffiths, Esq., at Campden, Gloucestershire, where it was bearing a good crop of cones near the summit in August 1905, and measured about 25 feet in height; this seems to show that the tree prefers limestone. At Kew there are three fine trees which were raised from seed obtained from Belgrade in 1889. These trees are now (1905) 13 inches in girth at 5 feet from the ground, and the tallest one is 23 feet high, making a strong, vigorous leading shoot, and assuming the very narrow pyramidal form which is so remarkable in the wild trees. The other two are 18 and 20 feet in height.

At Tortworth Court it has attained about 15 feet in height, and produced cones containing in the year 1902 apparently good seed; but Lord Ducie tells me that no plants raised from them can now be found. Though the tree is a very ornamental one I do not expect it can have any value as a forest tree in Great Britain, its timber having, so far as known, no special use. Judging from the soil and climate of its native country it should succeed in the Highlands of Scotland, especially on limestone soil, as well as, or better than in England, and as seedlings can now be procured in small numbers it will no doubt be planted by all lovers of coniferæ.   (H. J. E.)

PICEA BREWERIANA, BREWER'S SPRUCE

Picea Breweriana, Watson, *Proc. Amer. Acad.* xx. 378 (1885); Sargent, in *Gardeners' Chronicle*, xxv. 498, f. 93 (1886), and *Silva N. America*, xii. 51, t. 601 (1898); Kent, in Veitch's *Man. Coniferæ*, 430 (1900).

A tree, attaining 100-120 feet in height, with a stem 2 to 3 feet in diameter above its enlarged base. Branches crowded to the ground, with slender, pendulous branchlets, which are often 7 to 8 feet in length and sparsely covered in their first and second seasons with greyish pubescence. Pulvini long and slender, directed forwards. Leaves often nearly an inch long, rounded on both surfaces, the dorsal surface keeled and bearing 10 to 12 rows of stomata, the ventral surface dark green, shining with a prominent midrib, apex obtuse or short pointed. The leaves, on account of the shoots being pendulous, are radially arranged (never pseudo–distichous), their apices pointing outwards and downwards.

Cones on short stalks ($\frac{1}{4}$ inch), oblong-cylindrical, gradually narrowed from the middle to each end, $2\frac{1}{2}$ to 5 inches long by $\frac{3}{4}$ to 1 inch wide; scales broadly obovate with entire rounded margins; bracts minute, concealed, oblong, with denticulate upper margin. Seed with long wing (three times the length of the seed itself). The cones are pendent, greenish, or purplish green when fully grown, becoming dull brown when ripe, and open to let out the seed in autumn, but generally remain on the branches till the winter of the following year.   (A. H.)

This tree has a more limited range than any other spruce, being confined, so far as we know at present, to a few stations in northern California and south-western Oregon, on the Siskiyou Mountains, where it was discovered at an elevation of about 7000 feet, in June 1884, by Mr. Thomas Howell, who directed me to the best place from which the locality can be approached, a settlement called Waldo, about 40 miles west of Grant's Pass station, on the Southern Pacific Railway.

I went to this station in August 1904 with the intention of visiting Waldo; but finding that Messrs. Jack and Rehder, of the Arnold Arboretum, had just returned, and hearing from them that there were no cones on the trees in that year, I did not feel inclined to spend three days on the trip. I am, however, much indebted to these able botanists for the following information, and especially to Mr. Rehder for a beautiful negative of the tree, which is here reproduced. (Plate 29.)

There seems to be only a small grove of the trees about 20 miles south of Waldo, over the Californian boundary, which is best reached by following the trail to Happy Camp, and turning west near the summit of the pass to a place called Big Meadows, which is four miles from the pass.

There is another place where it grows near Selma, which is more accessible than Big Meadows, and other localities are mentioned by Sargent, who says that Professor Brewer, after whom the tree was named, had previously, in 1863, found a tree which was probably the same species, on Black Butte to the north of Strawberry Valley, at the western base of Mount Shasta, where, however, it cannot now be rediscovered.

Another locality for Brewer's spruce was found in 1898, by Mr. F. Anderson, on an unnamed but conspicuous peak at the headwaters of Elk Creek, about two or three miles west of Marble Mountain and eighty miles west of Mount Shasta. The elevation of the peak is about 8000 feet, and several hundred specimens were found growing near the summit; the trunks were 16 to 20 inches in diameter at 3 feet from the ground, and there were plenty of cones on the tops of the trees which were about 80 feet high.[1]

It grows on the Siskiyous in company with *Pinus ponderosa*, *P. Lambertiana*, and *P. monticola*, but usually gregariously in groves by itself. The soil and climate are dry, but there seems to be no special reason why this tree has proved in the eastern States of North America so difficult to cultivate; and as some of the conifers of the Pacific Coast which will not grow, or are not hardy in the eastern States, as, for instance, *Abies bracteata* and *Picea sitchensis*, thrive in England, and the trees with which it is associated in America are hardy and produce good seed here, we need not despair of seeing this beautiful tree established in the south of England.

The late Mr. R. Douglas, of Waukegan, Ill., visited Oregon in 1891 on purpose to obtain the seeds, and collected a large quantity of cones, from which several hundred thousand seedlings were grown. But those sown in America perished in their first and second years from causes which are not known, and attempts to raise the tree in the Arnold Arboretum have also failed.

Some of the seed, however, was raised by the late Baron von St. Paul Illaire at Fischbach in Silesia, which were alive in 1895;[2] and small plants were reported in 1903 to be growing in the Royal Pomological Institute at Proskau in Silesia.[3]

The late Mr. Johnson, of Astoria, Oregon, transplanted a few small trees to his nursery, some of which are, I believe, growing near Portland. Brandagee found a few two-year-old seedlings among the old trees, and half a dozen of them reached the Arnold Arboretum alive.

One of these was sent from there to Kew in November 1897, and is growing near the Pagoda, being about 2½ feet in height at the present time (March 1905). It is the only living specimen known to us in Britain.

The tree is said by Professor Sheldon to grow from 100 to 150 feet high, but Sargent gives 120 feet as the extreme height, and Messrs. Jack and Rehder did not see any higher than about 110 feet by about 9 feet in circumference. Douglas informed Baron von St. Paul that the largest tree measured by him was 121 feet high, and 2 feet 11 inches in diameter at 7½ feet from the ground. As the region in which it grows is so limited, and forest fires are very prevalent and

---

[1] *Erythea*, vi. 12 (1898), and vii. 176 (1899).   [2] *Mitt. Deutsch. Dendr. Ges.* 1895, p. 42.
[3] *Ibid.* 1903, p. 77.

destructive, it is to be feared that unless special measures are taken for its protection by the State of Oregon this very beautiful tree may become extinct.

The timber, which I only know from a specimen in the Jesup Collection of North American Woods, preserved in the American Museum of Natural History at New York, is said by Sargent to be considerably heavier than that of other American spruces, soft, close-grained, with a satiny surface, the sapwood hardly distinguishable.   The specimen alluded to is $13\frac{1}{4}$ inches in diameter under the bark at 166 years old.   (H. J. E.)

PICEA AJANENSIS, AJAN SPRUCE

Picea ajanensis, Fischer, *ex* Lindley and Gordon, *Trans. Hort. Soc.* v. 212 (1850), and in Middendorff,
  Reise, Florula Ochotensis, 87, tt. 22-24 (1856); Masters, *Jour. Linn. Soc. (Bot.)*, xviii. 508
  (1880), and *Gard. Chron.* 1880, xiii. 115, and xiv. 427, with figures; Mayr, *Monograph der
  Abietineen des Jap. Reiches*, 53, 102, t. iv. (1890); Kent, in Veitch's *Man. Coniferæ*, 425 (1900).
Picea ajanensis, var. *microsperma*, Masters, *Jour. Linn. Soc. (Bot.)*, xviii. 509 (1880).
Picea jezoensis, Carrière, *Traité Gén. Conif.* 255 (1855).
Abies ajanensis, Lindley and Gordon, *loc. cit.* (1850).
Abies jezoensis, Siebold et Zuccarini, *Flora Japonica*,[1] ii. 19, t. 110 (*ex parte*) (1844); Veitch, *Man.
  Coniferæ*, ed. 1, p. 72 (1881).

A tree, attaining in Yezo 100-150 feet in height. Bark like that of the common
European spruce, grey, and composed of irregularly quadrangular scales which do
not fall off. Branchlets shining, glabrous, yellow, never becoming reddish. Free
part of the pulvini long, directed backwards on branchlets of old trees, not widened
or channelled at their bases on the upper surface of the branchlets, persistent on
old branchlets. Buds broadly conic, with ovate scales rounded in margin, showing
on opening the young leaves tinged with red. Leaves flattened, thin, blunt, or ending
in a short point, slightly keeled on both surfaces; ventral surface green without
stomata; dorsal surface silvery white with two broad bands of stomata. Cones
purple when young, brownish when ripe, straight, oblong, tapering at each end,
2 to 3 inches long by nearly 1 inch wide; scales narrowly oblong-oval, coriaceous, erose,
and denticulate in margin; bracts minute, concealed, broad-oblong, slightly narrowed
below, their upper rounded denticulate edge giving off abruptly an apiculus. Seed
with a wing, which is twice or thrice as long as the seed itself.

IDENTIFICATION. (See *Picea hondoensis*.)

DISTRIBUTION

Picea ajanensis appears to be confined to Manchuria, Amurland, that
part of Eastern Siberia which faces the southern half of the Sea of Ochotsk,
Saghalien, the three southern isles of the Kurile group, and Yezo. The spruce of
Central China, which has been identified with it in *Index Floræ Sinensis*, ii. 553, has
pubescent shoots, and is probably identical with *Picea brachytila*, Masters. The
accounts of the Ajan spruce on the continent of Asia are of ancient date, the only
recent one being that in Russian by Komarov,[2] who states that it grows abundantly
with species of Abies and *Pinus koraiensis* in mountain woods in all the provinces
of Manchuria. It has not, however, been collected there by any British travellers.

---

[1] The figures given by Siebold represent (1) a flowering twig which came from a garden in Tokyo, and was probably,
according to Mayr, *Picea hondoensis*; and (2) a branch with cones, copied from a Japanese drawing of *Picea ajanensis* from
Yezo. The description applies to two species, and the name *jezoensis* cannot stand. The synonymy is very involved, but,
accepting Mayr's view, the facts are clear enough. The Hondo spruce was first distinguished clearly by Mayr, and therefore
receives his name *Picea hondoensis*. The Yezo and Amurland spruces are the same species, and receive the name *Picea
ajanensis*, first given by Fischer.

[2] Komarov, *Flora Manshuriæ*, i. 200 (1901).

Farther north, according to Maximowicz,[1] it extends throughout the territory of the lower Amur and the coast province facing the Sea of Ochotsk, reaching its northern limit in the interior in the Stanovoi mountains about latitude 55° 50′, and on the coast at Ajan, lat. 56° 27′. Schmidt[2] says that thick forests of *Picea ajanensis* occur in the lower Amur and in the coast territory. A mountain at 1000 feet in the Amgun valley was clothed with a thick mossy wood of this spruce, in the shadow of which snow still lay on the 30th May. On the crest of the Bureja range it occurs as a low prostrate shrub. It descends very seldom to the river banks. Middendorff also notes that it is confined to the hills on the coast of the Sea of Ochotsk. Occasionally it grows on swampy flats in Amurland.

Schmidt describes the bark as being moderately rough and divided into generally 6-angled plates, about an inch in diameter and ½ to 1 line in thickness; and that the form and colour of the leaves are very variable, their points being either acute or obtuse.

In the island of Saghalien, in its south-western part, there is a coniferous forest composed of *Picea ajanensis* and *Abies sachalinensis*, which clothes the slopes of the mountains up to 800 feet on the coast, and higher in the interior, where even the lofty crests are covered with dark forests of these two species.

In the Kurile Isles[3] this species is confined to the three islands north of Yezo, namely Kunashiri, Shikotan, and Etorofu, reaching its northern limit in the last named. In Shikotan it forms with *Abies sachalinensis* a dense mixed forest, which in habit and height and cover of the ground strikingly resembles the coniferous forests at moderate elevations in Germany. The cones borne by the tree in this island are, however, small in size, and the tree itself does not attain its maximum dimensions.

In Yezo, Mayr reports that he has seen trees 130 feet in height, and considers reliable the reports of the Japanese foresters that it occasionally attains even 160-200 feet. It occurs in all the mountains of Yezo, only reaching the coast in the west of the island, where it is found in cold, marshy localities immediately behind the dunes, being only separated from the sea by a growth of *Rosa rugosa* and shrubby *Quercus dentata*. The important forests of it lie in the western and central mountains of Yezo, and also in the high ranges of Kitami, Kushiro, and Nemoro, where it forms mixed woods with the Saghalien silver fir and *Picea Glehnii*.

INTRODUCTION

We do not know that any plants of the continental Ajan spruce have been grown in Europe.

John Gould Veitch visited Hakodate in 1860, and sent home specimens and seeds of a weakly form of the Yezo *Picea ajanensis*, which was described by Lindley[4] as a

---

[1] Maximowicz, *Primitiæ Floræ Amurensis*, 261, 392 (1859). See also Regel, *Tentamen Floræ Ussuriensis*, 136 (1861).
[2] Schmidt, "Reisen in Amurland und Saghalien," in *Mém. Acad. Imp. Sc. St. Petersburg*, VII. series, xii. No. 2, pp. 15, 20, 63, 98 (1868).
[3] Mayr, *loc. cit.* p. 102.
[4] *Gard. Chron.* 1861, p. 22. This is *Picea ajanensis*, var. *microsperma*, Masters, *Gard. Chron.* 1880, i. 115.

distinct species, *Abies microsperma*. Plants raised from the seed "turned out to be unsuitable for the climate of this country."[1] This form, according to Mayr, and so far as I can judge myself, can hardly rank even as a variety, and is not in cultivation at the present time.

Maries[2] visited Yezo in 1879 and sent home specimens, now preserved in the Kew Herbarium, and seeds of the true *Picea ajanensis* from that island; and young trees should accordingly be in cultivation in this country. This plant was kept separate by Messrs. Veitch at first, under the name *Abies yezoensis*. Maries considered the Yezo spruce to be quite distinct in habit and aspect from the two spruces which he had seen on Fujiyama (*Alcockiana* and *hondoensis*).

Mayr informed me last year that the Yezo spruce was not introduced into Europe until 1891; and that most of the trees on the Continent passing under the name of *Picea ajanensis* belong to *Picea hondoensis*. The specimens which have been sent me from old trees of reputed *P. ajanensis* in England also belong to that species. (A. H.)

On account of the heavy floods which occurred in July 1904, I did not get far enough north in Hokkaido to see this tree at its best, but in the State forests of Shari, Kutami, and Kushiro, it occurs in great masses, and is one of the principal economic products of the island. I saw it thinly scattered in forests of deciduous trees between Sapporo and Asahigawa, where it was of no great size, and in the forest round the volcanic crater-lake of Shikotsu in the south-east of Hokkaido it formed, here and there, nearly pure forests of small extent, mixed more or less with *Picea Glehnii* and *Abies sachalinensis*, at an elevation of 1000 to 2000 feet. The vegetation in these forests was quite unlike anything that I saw in Central Japan, the ground being covered with a dense layer of humus, and in the more shady places two or three species of *Pyrola* were abundant. *Daphne*, *Gaultheria*, *Ledum*, and other plants not seen elsewhere occurred, with curious terrestrial orchids and many ferns. The trees rarely exceeded 80 feet in height by 4 to 6 feet in girth, but higher up near the lake I measured one as much as 100 by 9 feet.

The general appearance of the tree is very like that of *P. sitchensis*, though I did not notice that the roots became buttressed, which is probably only the case in wet soil. The natural reproduction is good, but the seedlings grow slowly at first and seemed to thrive best in shade. The Japanese name is *Eso-Matsu*.

Timber

The wood of this tree is soft, but probably as good as that of other spruces. I passed the night at a factory in the forest where it was being cut up into thin slices for export to Osaka, where large quantities are used for making matchboxes. It is also employed for boat masts and other purposes, and is worth in Tokyo about 10d. per cubic foot. On account of its softness, lightness, and fineness of grain,

---

[1] Kent, in Veitch's *Man. Coniferæ, loc. cit.*
[2] See Veitch's *Man. Coniferæ*, ed. i. p. 72 (1881).

it is largely used in Japan for chip-braid, a peculiar Japanese industry, which has lately attained considerable importance, the export for 1903 amounting to no less than 1,363,000 yen—equal to about £140,000. This braid is mainly used for making hats and bonnets, but it is also woven into floor-matting, and as shown at the St. Louis Exhibition is both ornamental and cheap.

There are many different varieties of chip-braid, some of which are dyed of different colours, and others are plaited with a mixture of silk. It is exported in bundles of 50 to 60 yards long and 1 to 1½ inches wide, and is valued according to quality at 1s. to 6s. per bundle. The best are made by mixing chips of *Populus tremula* and *Picea ajanensis*. (H. J. E.)

PICEA HONDOENSIS, Hondo Spruce

Picea hondoensis, Mayr, *Monograph der Abietineen der Japanisches Reiches,* 51, t. iv. fig. 9 (1890) ;
Shirasawa, *Iconographie des Essences Forestières du Japon,* text 20, tab. v. figs. 1-22 (1900).
Picea ajanensis, Hooker, *Bot. Mag.* t. 6743 (1884), and of most writers.
Abies ajanensis, Fisch., var. *japonica,* Maximowicz, *Iter secundum* (1862).

A tree, attaining 80 feet in height in Hondo, the main island of Japan. Bark dark grey, peeling off in small roundish scales and leaving light-coloured spots on the trunk. Branchlets shining, glabrous, yellow in the first year ; but becoming reddish brown in the second year, and retaining the red colour in succeeding years till the scaly bark begins to form. The free portions of the pulvini are directed forwards, and on the upper side of the branchlets are enlarged transversely at their bases and show two channels where they become decurrent on the stem ; they are shorter than in *Picea ajanensis,* and on older branchlets tend to disappear. Buds like those of *Picea ajanensis,* but opening with greenish leaves. Leaves as in that species, but slightly shorter. Cones, red when young, yellowish when ripe, slightly curved, oblong, tapering to each end, about 2 inches long by ¾ inch thick, erect on terminal younger branchlets ; scales membranous, oval, broader proportionately to their length than in *P. ajanensis,* with denticulate erose margins ; bracts minute, concealed, oval lanceolate, denticulate, gradually tapering to an acute apex. Seed with a short wing (less than twice the length of the seed).

The description just given enumerates the characters, chiefly those of the bark, shoot, and cones, on which Mayr relies to distinguish the Hondo spruce from the true *Picea ajanensis.*

Picea hondoensis, as grown in this country, where it is usually called *Picea ajanensis,* assumes a broadly pyramidal outline, the main branches being rigid and directed either upwards or horizontally. In sunshine the branchlets turn their tips upwards, exposing to view the pale surface of the leaves. The arrangement of the leaves on lateral branchlets is the one normal in flat-leaved spruces, *i.e.* the upper side of the branchlet is densely covered with leaves, which have their apices directed forwards, while on the lower side of the branchlet the leaves part into two sets, directed outwards at right angles and leaving the twig bare beneath. All the leaves direct their stomatic pale surfaces away from the light, so that these look towards the ground.

The young cones are bright crimson, and make the tree highly ornamental in spring.

IDENTIFICATION

Picea Alcockiana, in which the leaves are conspicuously white on the dorsal surface, is often confounded in gardens with *Picea hondoensis* ; but these two species are readily distinguished as follows :—

Picea hondoensis.—Leaves flat, with bands of stomata confined to the dorsal surface. On the lower surface of lateral branchlets the twig is bare, with the leaves directed outwards at right angles.

Picea Alcockiana.—Leaves quadrangular in section, with lines of stomata on the ventral surface, in addition to the bands of stomata on the dorsal surface. On the lower surface of lateral branchlets the twig is not quite bare, and the leaves are directed forwards at an acute angle.

Picea hondoensis, Picea ajanensis, and *Picea sitchensis* have been distinguished, so far as leaves and branchlets are concerned, in the key to Section *Omorica.* The cones of these three species are much alike. Those of *Picea sitchensis,* however, have scales oblong in outline, with their upper edge scarcely emarginate or erose; the bracts are large and visible between the scales towards the base of the cone. In the other two species the scales of the cones are oval with erose margins, while the bracts are minute, concealed, and differently shaped.

The cones of *Picea Alcockiana* differ considerably from those of the three preceding species. Their scales are rounded, being nearly semicircular in outline, with the upper edge almost entire or only minutely denticulate; and their outer surface is markedly striated.

DISTRIBUTION

Picea hondoensis is confined to the central chain of mountains in the main island of Japan, occurring at altitudes above 4000 feet. Shirasawa (*loc. cit.*) mentions as localities, Fuji, Mitake, Novikura, Sirane to Nikko, Chokarsan to Ugo, etc.; and says that in the lower levels it is accompanied by *Tsuga diversifolia,* and ascends to 8000 feet in company with *Abies Veitchii.* Mayr states that on Fuji it is accompanied by *Picea bicolor* (*Alcockiana*), both occurring in mixed woods with *Larix leptolepis* and *Abies Veitchii.* Farther north, *Picea polita* joins the two spruces just named; and all three reach their northern limit in the high mountains of Iwashiro at 38° lat. Its southern limit is 35° lat.

Elwes saw very little of this tree in Japan, but near the top of the Wada-toge pass there were some small spruces growing at about 4500 feet elevation, which he believes to have been this species. *Tohi* is the Japanese name.

INTRODUCTION

Picea hondoensis was introduced in 1861 by John Gould Veitch. It was distributed as *Abies Alcoquiana,* an unfortunate circumstance, due to the fact that the seeds of the two spruces growing on Fujiyama (*Picea hondoensis* and *Alcockiana*) were both collected for Mr. Veitch by natives and were mixed together. Dr. Masters cleared up the question as to the distinctness of these two species in an article in the *Gardeners' Chronicle,*[1] in which, however, he retained the name *Picea ajanensis* for the spruce, which Mayr afterwards separated as *Picea hondoensis.* If

[1] *Gard. Chron.* 1880, i. 115, and ii. 427.

Mayr's view of the specific distinctness of *Picea hondoensis* and *Picea ajanensis* be upheld, most of the specimens cultivated in this country under the latter name (and many also incorrectly labelled *Alcockiana*) must be renamed as *Picea hondoensis*.

The best specimen we have seen in England is a tree at Hemsted in Kent, which was planted by the Earl of Cranbrook in 1887, and, when measured by Elwes in 1905, was 44 feet high.

There is one at Benmore, near Dunoon, the property of H. S. Younger, Esq., which Henry measured in 1905 as 52 feet by 4 feet 4 inches, about twenty-five years planted.

At Fota, Co. Cork, there is a fine tree which, in 1904, Henry found to be 44 feet by 4 feet 3 inches. (A. H.)

PICEA SITCHENSIS, Menzies' or Sitka Spruce [1]

Picea sitchensis, Carrière, *Traité Conifer.* 260 (1855); Trautvetter et Meyer, in Middendorff, *Reise*
    Florula ochotensis, 87 (1856);[2] Sargent, *Silva N. America*, xii. 55, t. 602 (1898); Kent, in
    Veitch's *Man. Coniferæ*, 452 (1900).
Picea Menziesii, Carrière, *Traité Conifer.* 237 (1855); Masters, *Gard. Chron.* xxv. 728, figs. 161,
    162 (1886).
Picea sitkaensis, Mayr, *Wald. N. Amerika*, 338 (1890).
Pinus sitchensis, Bongard, *Vég. Sitcha*, 46 (1832).
Abies Menziesii, Lindley, *Penny Cycl.* i. 32 (1833); Loudon, *Arb. et Frut. Brit.* iv. 2321 (1838).
Abies sitchensis, Lindley and Gordon, *Jour. Hort. Soc.* v. 212 (1850.)

A tree, sometimes exceeding 200 feet in height, with a trunk 4 to 20 feet in
diameter, tapering above its enlarged and buttressed base; in Alaska dwindling to
a low shrub. Bark with large, thin, red-brown scales. Branchlets yellow, shining,
glabrous. Buds ovoid, acute at the apex, with ovate obtuse scales. Leaves
arranged on lateral branchlets as in *Picea ajanensis*, ending in sharp cartilaginous
points; deeply keeled on the ventral green surface, and almost convex on the dorsal
surface, which has two white broad bands of stomata. The male catkins are solitary
at or near the ends of the branchlets, and are of an orange reddish colour.

Cones: on short straight stalks, cylindrical-oval, blunt at the free end, $2\frac{1}{2}$ to 4
inches long by 1 to $1\frac{1}{2}$ inches wide, composed of oblong or oblong-oval scales,
rounded towards the apex, denticulate and scarcely erose in margin; bracts lanceolate,
denticulate, about half as long as the scales, and peeping out between them towards
the base of the cone. The cones when ripe are yellow or brown, and generally fall
off in the autumn and winter of the first year. Seeds, with a wing, three to four
times as long as the seed itself.

The Sitka spruce seems to vary considerably over its wide area. There are
specimens at Kew from the Columbia River, with pubescent young shoots, and
bearing small cones which have oval, not oblong, scales, and minute almost orbicular
bracts. Other specimens from Alaska have larger cones than usual, but with bracts
shorter than usual, and the leaves are not so deeply keeled or so sharp-pointed as in
the type.

Cultivated trees are generally broadly pyramidal in outline, and when old,
often show the enlarged and buttressed base, so characteristic of wild trees; the roots
sometimes extending superficially above the ground for several feet. The tree
often produces on its lateral branches small erect shoots, on which the leaves spread
radially in all directions.   (A. H.)

IDENTIFICATION.   (See *Picea hondoensis*)

---

[1] Called also Tideland spruce on the Pacific coast.
[2] Trautvetter and Meyer are often cited as the authors of the name *Picea sitchensis*; but the correct date of their
publication is later than that of Carrière's. See Trautvetter, *Floræ Rossicæ Fontes*, 303 (1880).

Varieties

On the Continent, according both to Beissner[1] and to the late Prof. Carl Hansen, whose "Pinetum Danicum," published in the *Journal of the Royal Horticultural Society for* 1892, is a valuable contribution to our literature, a variety (*speciosa*) occurs in cultivation, which is light blue in colour and very decorative. It differs from the ordinary form in being slower in growth and in having leaves, which are shorter, stiffer, and more sharply pointed.

Distribution

According to Sargent, this spruce extends farther north-west than any other North American conifer, being found in long. 151° west on the east end of Kadiak island, and all through the coast region of Alaska and British Columbia, west Washington and Oregon, and as far south as Caspar in Mendocino County, California.

In the north it is a small tree, sometimes only a bush, but on the coast of south-east Alaska is the largest and most abundant tree, and grows in company with the western hemlock. Here it attains over 100 feet in height, and ascends the mountains to about 3000 feet.

In the south of British Columbia it is larger in size, but in Vancouver's Island it did not seem common, and was not a conspicuous tree in the south-east parts of the island which I visited.

In Washington it grows to a very large size, and I measured one in swampy ground near a logging camp in the White River valley which was 23 feet in girth at 6 feet from the ground, and appeared to be 3 to 4 feet in diameter at the place where it was broken off at about 120 feet from the ground.

Prof. Sheldon, in a pamphlet on *The Forest Wealth of Oregon*, calls it the largest tree in the state, growing 200 to 300 feet high, and has figured as the frontispiece of this paper what he calls the largest Tideland spruce in the world. This tree grew on the coast in God's Valley, on the North Nehalem River, Clatsop County, Oregon, and measured 30 feet 11½ inches in diameter at 2 feet from the ground, and 20 feet 4½ inches at 6 feet from the ground.

He states that it is distinctly a moisture-loving tree, and in the extensive coast belt forest which it forms is an ideal lumber tree, free from limbs for a great part of its height.

It is not mentioned as growing in the great forest reserve of the Cascade Range, and, according to Sheldon, extends southwards along the coast as far as Curry County. In northern California it grows on rich alluvial plains at the mouths of rivers, or in low valleys facing the ocean, where it is associated with *Sequoia sempervirens* and *Abies grandis*, and thus may be said to be almost strictly confined to a region where there is perennial moisture in the air, and an annual rainfall of 50 inches and upwards.

[1] *Nadelholzkunde*, 392 (1901).

Sargent says its growth is very rapid, the leading shoots of young trees on Puget Sound being often 3 to 4 feet long.

John Muir measured a tree in Washington 180 feet high, at 240 years old, with a trunk 4 feet 6 inches diameter. Another near Vancouver, B.C., only 48 years old, had a trunk 3 feet in diameter.

In Alaska, Gorman measured two trees—one grown in a dense wood, well protected from wind, was 160 feet high, at 267 years old, with a diameter of 3 feet 11 inches; and the other on a hillside exposed to fierce north-east gales, was $4\frac{1}{2}$ feet in diameter at 14 feet from the ground, and 434 years old. The heart of this tree was 32 inches from the south-west side, and only $16\frac{1}{2}$ inches from the north-east side, showing the effect of prevalent winds on the production of branches and wood.

A tree measured by Muir at Wrangel, Alaska, was no less than 764 years old, with a trunk 5 feet in diameter, and this, I think, is the greatest age to which any recorded spruce has attained.

INTRODUCTION

Though discovered in Puget Sound in May 1792 by Archibald Menzies, who was surgeon and naturalist to Vancouver's expedition, it was not introduced to cultivation until 1831 by David Douglas, and was described by Lindley under the name of *Abies Menziesi* one year after Bongard had made it known to science under the specific name which we adopt. It is, however, still commonly known in Great Britain as Menzies' spruce, and his name it may well bear. According to Loudon, only a very few plants were raised in the Horticultural Society's Garden in the year 1832, of which some still survive.

The Oregon Association, which was formed a little later by a few Scottish arboriculturists for the purpose of introducing the conifers of the Pacific coast, and who sent out John Jeffrey as a collector about 1850, were fortunate in procuring a large quantity of seed, from which the pineta of Scotland and England have been stocked, and it has now become a common tree.

CULTIVATION

Though Menzies' spruce loves a wet climate, it loves a wet soil even more, and soon becomes unsightly and loses its foliage in dry localities. No conifer, except perhaps the Douglas fir, grows so rapidly where it has a suitable situation, and in some parts of Scotland it is now being planted experimentally as a forest tree.

It is easily raised, either from home-grown or imported seed, and is, like all spruces, slow of growth for the first few years, and requires at least five or six in the nursery before it is large enough to plant out.

At Durris, in Aberdeenshire, on the property of Mr. H. R. Baird, there is a plantation of Sitka spruce about 15 acres in extent, of which Mr. John Crozier, the forester in charge of the estate, gives us the following particulars in a letter dated 12th September 1904 :—

"The plantation occurs at an altitude of 700 to 800 feet, the aspect being northerly, soil a sandy peat over boulder clay. The age of the plantation is twenty-five years; but there is no record either as to the number of the plants put out or the age when planted. They were, however, notched in, and their age would most probably be four years. They were planted rather irregularly, the distance varying from 6 to 9 feet, and both common spruce and Scots fir have been used to fill up between, to 3 feet between each plant. The average height of the Sitka trees is about 33 feet; and the girth at 5 feet taken at random is (where they had been planted 9 feet apart), 24 in., 22 in., 20 in., 25 in., 22 in., 22 in., 28 in., 22 in., 26 in., 25 in. The *largest* common spruces I could find on the same ground measured 9 in., 8 in., 11 in., 9 in., 11 in., 12 in., 16 in., 9 in., 10 in., 12 in., and their height was about 26 feet. I took the measurement of a hundred Sitkas over a track 20 feet broad, just as they came, and they averaged 22¾ in. Where the Sitkas had been planted at 6 feet apart, the common spruce and Scots fir are dominated, and the greater part of them quite dead. I drained some very wet parts a year ago, where both the Scots fir and common spruce had been killed through excess of moisture, but the Sitka had been very little harmed by it. Judging by what I have seen of the tree here and elsewhere, it will stand a greater degree of moisture than any other conifer I know. The plantation is altogether in a very healthy state."

A few hundred Menzies' spruces were planted out in 1879 on the mountain at Bronydd, on the property of Lord Penrhyn in North Wales, at 900 feet elevation; according to Mr. Richards, the forester, only half a dozen trees now survive, in a wretched condition. He states, however, that as the young growths come out late in the spring the tree is never touched by frost in North Wales. At Penrhyn there is a good specimen of the tree measuring 10 feet 6 inches in girth in 1904.

Menzies' spruce,[1] on account of its sharp needles, has been supposed to be free from the attacks of deer, rabbits, and hares; but recent observations made in the royal domain of Freyr in Belgium show that out of 10,000 plants introduced some years ago only 2000 remain, and these are not expected to survive long. This is much to be regretted, as they had grown splendidly.

REMARKABLE TREES

One of the largest trees we know of in the south of England is at Highclere, Berks, the seat of the Earl of Carnarvon, where we measured a tree in August 1903 which was 96 feet by 12 feet. The tree, having lost its lower branches owing to a heavy snow-storm, has put out new branches down the trunk, a somewhat rare occurrence in large conifers. Another very fine tree is growing at Barton, Suffolk, which was planted in 1847, and when measured by Henry in 1904 was 99 feet by 9 feet 3 inches. Both of these are in a dry climate but in a good soil.

At Bicton, Devonshire, I measured a tree in 1902 which was 85 feet by 11 feet

---

[1] *Bull. Soc. Cent. Forest. Belgique*, April 1901.

6 inches; and at Boconnoc, Cornwall, the seat of J. B. Fortescue, Esq., a tree was recorded in 1891 as being 85 feet high by 12 feet in girth at the age of 48 years.[1]   In 1905 Elwes measured this and found it to be 86 by 15 feet.

At the same time[1] a tree growing at Howick Hall, Northumberland, the seat of Earl Grey, was 90 feet high at the age of 58 years.

At Beauport, Sussex, a tree measured in 1904 95 feet by 12 feet 10 inches.   It has very wide-spreading superficial roots, one extending over the ground 16 feet in length.   According to Sir Archibald Lamb, the tree five years ago was 12 feet 3 inches at 3 feet up, its present girth (1904) at that height being 13 feet 4 inches. (Plate 30.)

In Scotland the largest tree we know of, and probably the largest in Great Britain, is at Castle Menzies, said to have been 46 years old in October 1892, when its exact measurement was given by Mr. J. Ewing as 96½ feet high by 11 feet in girth.   I measured it carefully in April 1904, and found it to be 110 feet high by 13 feet 2 inches. This tree is growing on the banks of a pond in good and damp soil, and has produced a greater amount of timber in a short time than any conifer I know in Scotland, except, perhaps, the Douglas fir, though *Sequoia sempervirens* may run it close in England. But spruce timber grown so fast is very soft, coarse, and knotty, close planting being essential to give the tree any economic value.

At Abercairney, Perthshire, there is a tree which was measured by Henry in August 1904, as 99 feet in height by 9 feet 9 inches in girth.   This was 76 feet by 7 feet 5 inches in 1891.[2]

At the Keillour Pinetum,[3] in the same county, on boggy ground on a hillside, there is a remarkable Menzies' spruce, 86 feet in height by 15 feet 9 inches in girth. It has wide-spreading buttressed roots, and is branched to the ground.   According to a MS. account in the possession of Col. Smythe of Methven Castle this tree was planted in 1834 or 1835.   In this pinetum many species of conifers were planted in these two years, and owing to the wet, boggy nature of the soil some kinds have grown slowly, such as *Picea nigra*, *Pinus Cembra*, *Abies balsamea*, and *Abies Pinsapo*.   *Abies grandis* has perhaps succeeded best, next to *Picea sitchensis*, which has produced an amount of timber far in excess of the other species.   *Abies grandis* here is 90 feet by 7 feet 3 inches.   A *Picea alba*, planted presumably at the same time, is only 52 feet by 5½ feet.

Mr. Crozier reports that there is a Menzies' spruce 13 or 14 feet in girth at Dunrobin in Sutherland.

At Murraythwaite, in Dumfriesshire, the seat of W. Murray, Esq., a tree 78 feet by 8 feet 10 inches, planted about the year 1855, is growing near a pond, and is a fine healthy specimen, broadly pyramidal, and feathered to the ground.

A tree[4] at Keir, near Dunblane, measured, in 1905, 82 feet by 9 feet 10 inches. At Smeaton Hepburn, Haddington, the seat of Sir Archibald Buchan Hepburn, Bart., where there is a remarkably varied collection of trees, a fine Menzies' spruce

---

[1] *Jour. Roy. Hort. Soc.* 1892, xiv. 486, 493.   [2] *Ibid.* 1892, xiv. 527.
[3] Visited by Henry in August 1904.
[4] This tree was reported in 1891 to be 61 feet by 7 feet 3 inches, and was then forty years old, *Jour. Roy. Hort. Soc.* (1892) xiv. 531.

measured, in 1905, 88 feet by 10 feet 7 inches. A very large tree is reported to be growing in the grounds of Major Ross at Kilroch, Nairnshire.

In Ireland the finest example that we know is at Curraghmore, Waterford, the seat of the Marquess of Waterford. Mr. Crombie writes that it is now (March 1905) 106 feet in height, with a girth of 12 feet at 5 feet from the ground. This tree was reported[1] in 1891 to be 110 feet high (evidently an estimate) with a girth of 10 feet.

At Mount Shannon, Co. Limerick, there is growing a very vigorous tree, with branches to the ground, which in 1905 was 79 feet by 12 feet.

A tree at Clonbrock, Co. Galway, planted in 1881 and growing in boggy soil, was in 1904 56 feet high by 4 feet 8 inches in girth.

TIMBER

The wood is said by Sargent to be light, soft, and straight grained, not strong, with a satiny surface, and thick, nearly white, sapwood.

It is largely used on Puget Sound for purposes where cheap lumber is required, but I did not see it in the timber yards that I visited in Tacoma.

Laslett does not mention it in his work, but Stone, quoting Macoun, says that it is elastic, bends with the grain without splitting, and is much used in boat-building, for light oars, staves, doors, and window-sashes, resists decay for a long time, and is not attacked by insects.

I am informed by Mr. Rogers, one of the principal timber buyers for the Admiralty, that no other spruce makes such good light oars, and that in consequence it is now imported annually for that purpose.   (H. J. E.)

[1] *Jour. Roy. Hort. Soc.* (1892), xiv. 562.

TAXUS

Taxus, Linnæus, *Gen. Pl.* 312 (1737); Bentham et Hooker, *Gen. Pl.* iii. 431 (1880); Masters, *Jour. Linn. Soc. (Bot.)* xxx. 7 (1893); Pilger, in Engler, *Pflanzenreich*, iv. 5, *Taxaceæ*, 110 (1903).

EVERGREEN trees or shrubs belonging to the division Taxaceæ of the order Coniferæ. Bark reddish or reddish brown, thin and scaly. Branches spreading, giving off branchlets, of one kind only, irregularly alternate, surrounded at their bases by brownish scales. Buds globular or ovoid, of imbricated scales. Leaves inserted on the branchlets in a spiral order, on upright shoots spreading radially, on horizontal shoots disposed by twisting on their petioles in one plane in a pectinate arrangement, the upper and lower leaves being of the same length, with their dorsal surfaces turned upwards and their ventral surfaces downwards. In fastigiate varieties all, or most, of the branchlets assume an erect position, and the leaves in consequence are arranged radially. The leaves are linear, flat, with recurved margins, dark green above, paler green below; the lower surface only bearing stomata, which never form conspicuous white bands; narrowed at the base into a short petiole, arising from a linear cushion on the twig; mucronate or acute at the apex and without a resin-canal.

Flowers diœcious, or in rare individuals monœcious, on the under surface of the branchlets of the preceding year, in the axils of the leaves, the female flowers being less numerous than the male flowers. Male flowers composed of a stalk, girt at its base by imbricated scales, and bearing above a globose head of 6-14 stamens with short filaments. The stamen is expanded above into a peltate connective, which bears on its lower surface 5 to 9 pollen sacs, united with each other and with the filament. The female flowering shoot, arising out of the axil of the leaf, is composed of a number of imbricated scales, in the axil of the uppermost one of which is borne an ovule, placed so close to the apex of the shoot as to appear terminal; in the scale next below a bud occurs, which occasionally develops into a second ovule. The ovule, which has a small membranous disc at its base, projects out of the scales by its micropyle. Seed sessile in a fleshy, juicy cup, forming an aril (the enlarged disc), open at the top and free from the seed in its upper part. The seed variable in form, 2, 3, or 4-angled, is generally ellipsoid and has a ligneous testa, containing oily white albumen, in the upper part of which is an axile straight cylindrical minute embryo with two cotyledons.

Yews differ from all other Coniferæ in the character of the fruit. They resemble in foliage certain other genera of Taxaceæ, but are readily distinguishable as follows :—

Taxus.—Branchlets standing irregularly alternate on the branches. Leaves stalked, greenish underneath with no definite bands of stomata. Buds composed of imbricated scales.

Pruminopitys.—Branchlets and leaves as in Taxus, but with valvate bud-scales.

Cephalotaxus.—Branchlets opposite. Leaves like the yew in consistence, but with white bands beneath showing definite lines of stomata.

Torreya.—Branchlets sub-opposite. Leaves rigid and spine-pointed with white bands beneath, showing definite lines of stomata.

Saxegothæa.—Branchlets in whorls, ascending at an angle. Leaves with bases decurrent on the shoots, and with white bands beneath which are narrow and close to the median line.

The genus is widely distributed over large parts of North America, Europe, Algeria, and Asia, and occurs sporadically in the mountains of Sumatra, Celebes, and the Philippines. Seven distinct species have been described, each confined to a definite territory. These species are, however, rather geographical forms, only differing from one another in trivial characters of foliage and habit. The view taken by Sir Joseph Hooker[1] and by Pilger, the latest monographer of the genus, that they only constitute one species is probably correct. Many of the supposed specific distinctions, such as the density of the foliage on the branchlets, the size and form of the leaf, etc., are due in most instances to the influence of soil, shade, and climate. Moreover, in the varieties of the common yew, which are known to have arisen as sports in the wild state or in cultivation, greater differences occur in the characters of habit, foliage, and fruit, than are observable in the so-called species. In the account which follows, the geographical forms will be treated as varieties.

[1] *Himalayan Journals*, ii. 25 (1854), and *Student's Flora of Brit. Islands*, 369 (1878).

TAXUS BACCATA, Yew

Taxus baccata, Linnæus, *Sp. Pl.* 1040 (1753); Loudon, *Arb. et Frut. Brit.* iv. 2066 (1838); Lowe, *Yew Trees of Great Britain and Ireland* (1897); Kent, in Veitch's *Man. Coniferæ*, 126 (1900); Kirchner, Loew, and Schröter, *Lebengesch. Blutenpfl. Mitteleuropas*, i. 61 (1904).

The chief characters of the species have been given in the generic description. The different geographical forms are distinguished as follows :—

1. Var. *typica*, Common Yew.—A tree or shrub. Leaves falcate, acute, or acuminate, the apex diminishing gradually into a cartilaginous mucro; median nerve only slightly prominent above. Buds ovoid or globose, of closely imbricated brownish, rounded scales, usually not keeled on the back.

In certain Himalayan specimens the leaves are long and narrow, with a long acuminate apex, and the buds have keeled scales. Intermediate forms occur; and all Indian botanists and foresters seem to be agreed that the Himalayan yew cannot be separated from the European form even as a variety.[1]

2. Var. *cuspidata* (*Taxus cuspidata*, S. et Z.[2]), Japanese Yew.—A tree or shrub. Leaves straight, scarcely falcate, median nerve prominent above, apex giving off abruptly a short mucro. Buds oblong, composed of somewhat loosely imbricated scales, which are ovate, very acute and keeled. In cultivated specimens the under surface of the leaves is yellow in colour, the buds being bright chestnut brown.

3. Var. *sinensis*,[3] Chinese Yew.—A tree. Leaves short, rigid, median nerve not prominent above, apex rounded and giving off abruptly a short mucro. Buds ovoid, brownish, composed of densely imbricated scales, which are ovate, obtuse, and not keeled.

4. Var. *brevifolia* (*Taxus brevifolia*, Nutt.[4]), Pacific Coast Yew.—A tree. Leaves falcate, short, median nerve slightly prominent above, apex abruptly mucronate. Buds large, with loosely imbricated yellowish green scales, which are lanceolate, mucronate, and keeled.

5. Var. *canadensis* (*Taxus canadensis*, Marshall[5]), Canadian Yew.—A low, prostrate shrub. Leaves narrow, falcate; median nerve slightly prominent above, apex abruptly mucronate. Buds globose, small, with somewhat loosely imbricated, greenish, ovate, obtuse, keeled scales.

6. Var. *Floridana* (*Taxus floridana*, Chapman[6]), Florida Yew.—A shrub or

---

[1] It has been described as a distinct species, *Taxus Wallichiana*, Zuccarini, in *Abhand. K. Bayr. Akad. Wissensch.* iii. 803, t. 5 (1843). Pilger, who ranks the different geographical forms as sub-species, keeps it separate from the European yew as sub-species *Wallichiana*.

[2] *Flora Jap. Fam. Nat.* ii. 108 (1846); Shirasawa, *Icon. Ess. Forest. Japon.* i. 33, t. 15 (1899).

[3] *Taxus baccata*, L., Masters, *Index Floræ Sinensis*, ii. 546 (1902).

[4] Nuttall, *Sylv.* iii. 86, t. 108 (1849); Sargent, *Silva N. America*, x. 65, t. 514 (1896).

[5] Marshall, *Arb. Amer.* 151 (1785); Sargent, *Silva N. America*, x. 63 (1896). The plant cultivated at Kew as *Taxus canadensis*, var. *aurea*, a strong-growing, erect shrub, is apparently a variety of the common yew.

[6] Chapman, *Flora South United States*, 436 (1860); Sargent, *Silva N. America*, x. 67, t. 515 (1896).

very small tree. Leaves very narrow, median nerve scarcely prominent, apex acute and gradually passing into the mucro. Buds small, with loosely imbricated, ovate, obtuse scales.

7. Var. *globosa* (*Taxus globosa*, Schl.[1]), Mexican Yew.—A small tree. Leaves variable, narrow, straight, acuminate, mucronate. Buds of numerous ovate, rounded, obtuse, keeled scales.

DISTRIBUTION

I. COMMON YEW.—All authorities are agreed that the yew was formerly much more widely spread in Europe than is the case to-day. Conwentz[2] relies on three points to prove the ancient wider distribution :—(1) fossil remains ; (2) prehistoric and historic antiquities ; (3) place - names. He considers that nearly all the fossil remains of the Tertiary age, which have been described as species of *Taxus*, are not really yew. In more recent geographical strata, however, numerous fossil remains of yew have been found. Clement Reid[3] gives the following list of deposits in which yew occurs in England :—

Neolithic.—Common in peat below the sea-level in the Thames valley and Fenland ; Portobello, near Edinburgh.

Interglacial.—Hoxne, Suffolk.

Preglacial (Cromer Forest-bed).—Mundesley, Bacton, Happisburgh (in Norfolk), Pakefield (in Suffolk).

Conwentz has found fossil remains in numerous localities in England and Ireland ; but his promised paper on the subject has not yet been published. Guided by place-names in Germany, he dug up fossil yew in many localities in that country.

He[4] found under pure peat, 3 feet thick, in the Steller Moss not far from Hanover, some hundreds of stems of yews. He says that it is never found in the ramparts of prehistoric forts, but that it was often planted on fortifications by the knights of the Middle Ages.

He has prepared a list of some hundreds of English, Scottish, and especially Irish names of places taken from the yew. The Gaelic name for the yew is *iubhar* ; and in Irish and Scottish place-names this generally appears Anglicised as *ure*, being sometimes corrupted into *o* or *u* simply. Youghal means yew-wood. Dromanure and Knockanure signify yew-hill. Glenure is the yew-glen. Gortinure and Mayo mean yew-field.

Conwentz examined prehistoric wooden boxes, buckets, and other vessels in the British Museum and in the Dublin Science and Art Museum, and identified the wood of some thirty articles as that of yew.

Yew is occasionally found in peat-mosses in Ireland, but is exceedingly rare as compared with pine and oak. Mr. R. D. Cole, who has kindly sent me a

---

[1] Schlechtendal, *Linnæa*, xii. 496 (1838) ; Sargent, *Silva N. America*, x. 63 (1896).
[2] *British Association Report*, 1901, p. 839.   [3] *Origin British Flora*, 151 (1899).
[4] *Bot. Centralblatt*, 1896, lxvi. 105 ; and 1900, Beihefte, ix. 223.

specimen of bog-yew, drew attention in 1903 to the occurrence of yew in Ballyfin bog in Queen's County. It was so plentiful there in former times that the farmers in the neighbourhood used it for gate-posts and in roofing houses. Mr. J. Adams has published a short account[1] of Mr. Cole's discovery, from which it appears that the cross-section of one trunk, 2 feet in diameter, showed no less than 395 annual rings. Another specimen showed 123 rings, only occupying a width of 1½ inches. Mr. Cole informs me that in no case where the root was vertical did he find more than 18 inches deep of peat beneath; in other parts of the bog where the yews were found more deeply buried, their roots were twisted and out of their natural position, and were probably carried there by floods. Apparently then, the yew, unlike the common pine, never grew in any great depth of peat.

Large trunks of yew were formerly dug up on the shore of Magilligan in Co. Derry, between the rocks and the sea.[2] On the east side of Glenveigh, in Co. Donegal, thick logs are reported to be often found in the peat.[3]

In the Kew Museum there is a specimen of fossil yew, which was dug up in Hatfield Chase, near Thorne, Yorkshire, from under a bed of clay 6 feet in thickness; and another specimen is labelled, "Submarine Forest, Stogursey, Somersetshire."

Professor Seeley, F.R.S., in a very interesting letter, dated January 1904, says that he has seen "the broken stumps of yew trees standing as they grew by scores, possibly by hundreds, in Mildenhall Fen, about 1865, when the peat was entirely removed so as to prepare the land for corn. One tree sketched by Mr. Marshall, at that time Coroner for the Isle of Ely, from a section between Ely and Downham Market, showed the yew growing in sandy gravel with black flints. The roots were entirely in the gravel. Above the gravel is the 'Buttery Clay,' 2 feet 6 inches thick, into which the trunk of the tree extended vertically, rising about 1 foot into the Upper Peat, which was 4 feet 6 inches thick. This clay is marine, and is the delta mud of the Cam and the Ouse deposited on the Lower Peat and beyond it, where a depression of land admitted the sea over the Isle of Ely and killed the forests. A little part of the Scrobicularia Clay is 6 feet thick, and the peat above it 18 feet thick. The common trees in the peat there are pines and oaks. I have never seen the beech, and never heard of the lime. About the pine there is no doubt. It occurred in the forests of the Forest Bed of Norfolk, and at several localities in the peat of the fens, almost always on clay covered by peat."

In the present day the common yew is met with growing wild in most parts of Europe, from Scandinavia to the Mediterranean, and from the Atlantic to the western provinces of Russia. It has only recently become extinct in the Azores. It also occurs in Algeria, Asia Minor, the Caucasus, North Persia, the Himalayas, and Burma. The yew also extends into the mountains of Sumatra, South Celebes, and the Philippines.[4]

---

[1] *Irish Naturalist*, xiv. 1905, p. 34, with plate showing yew trunk and transverse section.
[2] Mackay, *Flora Hibernica*, 260 (1836).
[3] Hart, *Flora County Donegal*, 237 (1898).
[4] Specimens from these localities have been identified by Pilger as the Himalayan yew.

In England the yew is indigenous on all the chalky Downs of Sussex, Hampshire, and Wilts. According to Bromfield,[1] the yew is one of the few natural ornaments of our South Downs, over the bare sides and summits of which it is scattered abundantly as single trees, frequently of great size and antiquity; sometimes in groups; more rarely forming groves in the bottoms or valleys between these rounded hills, or in the steep woods which clothe their sheltered slopes. He mentions as one of the most remarkable of these yew groves, that at Kingsley Bottom, near Chichester. The yew is remarkably plentiful on the banks of the Wye, about Chepstow and Tintern, and grows in the most inaccessible positions on the limestone cliffs there, as it does also on the rocks of Matlock. The rocks at Borrodale and on Conzie Scar, near Kendal, are also truly natural stations of the yew.[2] The yew is frequent in the woods of Monmouthshire, and in the ancient forest of Cranbourne Chase in Dorsetshire.[3] In the Wyre Forest it is certainly wild, occurring now as isolated trees amidst the beech and oak. In Seckley Wood, on the Severn, there are indigenous yew trees, one of which is remarkable for its curious pendulous habit.[4] It ascends to 1500 feet in Northumberland.[5]

Concerning the occurrence of the yew as a wild plant in Scotland our information is scanty. Hooker[5] states that it is indigenous as far north as Aberdeen and Argyll. White[6] records it from Breadalbane in Perthshire. Lightfoot,[7] writing in 1777, says it was found here and there in the Highlands in a truly wild state, and that there were the remains of an old wood of yew at Glenure in Upper Lorn, Argyllshire.

It is now of rare occurrence in the wild state in Ireland. According to Praeger,[8] it is found on rocks, cliffs, in old woods, and on lake shores, now almost confined to the west. It is recorded from various localities from Kerry to Donegal, and Praeger considers that some of these instances may represent the last remnants of aboriginal stock; but it is impossible now to say definitely, as introduced trees grow around the supposed wild ones. The yews in the rough wood at Avondale, in Wicklow, may be wild. Many years ago Moore[9] found the yew growing at Benyevena, in Co. Derry, in the crevices of the rocks, at an elevation of 1200 feet, when it assumed the appearance of a low shrub. In Smith's *Kerry* (1756), it is stated that "the yew grew in prodigious quantities in all our southern baronies until it was destroyed for making coals for the iron-works."[10]

In Norway the yew is called "Barlind," and, according to Schubeler,[11] grows wild only in the south, especially along the coast, the farthest point north known to him being near Sondmore, in lat. 62° 30″ N., where it attains the height of 32 feet. In the east it does not extend farther north than Hurdalen, lat. 60° 35′, where it attains a height of 8-10 feet. Schubeler mentions as the largest yews known

[1] *Flora Vectensis*, p. 472. [2] Lowe, *loc. cit.* p. 28.
[3] Strangways in Loudon's *Gard. Mag.* 1839, p. 119. [4] *Trans. Worcester Nat. Hist. Club*, 1847-1896, p. 16.
[5] *Stud. Flora Brit. Islands*, 369 (1878). [6] *Flora of Perthshire*, 283 (1898).
[7] *Flora Scotica*, ii. 626 (1777). [8] *Proc. Roy. Irish Acad.* vii. 290 (1901).
[9] Mackay, *Flora Hibernica*, 260 (1836). [10] *Cybele Hibernica*, 331 (1898).
[11] Schubeler, *Viridarium Norvegicum*, p. 448.

to him in Norway some at Tufte, on the Christiana fjord, which are 42-43 feet high, with a girth of 3 feet 4 inches to 3 feet 9 inches. The thickest one was, however, 4 feet 10 inches at 2 feet from the ground. He figures (p. 458, fig. 84) what is very rarely seen in England, a self-layered yew, and says that he found in a wood at Hallangen a tree 24 feet in length with a diameter of only 6 inches.

In Sweden the yew grows as far north as lat. 63° 10′, and thrives so well that a tree at Maltesholm, in Scania, is said to have had a diameter of 89 centimetres when only 75 years old. It occurs on the Swedish Island of Åland (lat. 60°), but only as a small shrub.

Its northerly limit in Russia appears to be Esthonia, its eastern limit also passing through that province, and continuing southwards through Livonia, Courland, Lithuania, Volhynia, Podolia, and the Crimea. It occurs also in Denmark,[1] but only in one place wild, viz., at Munkehjerg, the beautifully situated hotel near the town of Veile, in Jutland. Formerly the yew was much more widely spread in Denmark, but owing to the value of the wood the wild trees have been destroyed in most parts of the country.

In Belgium, where the yew is often planted, its occurrence in the wild state has been denied by some authors. Wildeman and Durand,[2] however, consider that it is probably wild in the neighbourhood of Huy and in Hainault.

In France[3] it occurs chiefly in mountainous regions, as in the Vosges (where it is rare), Jura, Cevennes, Pyrenees, and Corsica. In the Pyrenees it ascends to 5400 feet, and, according to Bubani,[4] is always rare (due to destruction by human agency), and only occurs on limestone and in cool and shaded situations. In France generally, it is most common on precipices and rocky spots, and nearly always on limestone. It never forms pure woods; but is, however, remarkably abundant in the forest of Sainte Baume (Department of Var), where the oldest and largest wild yew trees in France occur, some attaining a girth of 11½ feet. In Normandy, according to Gadeau de Kerville,[5] it is not indigenous, being probably introduced at a very early period before the conquest of Gaul by Julius Cæsar. It is usually planted in churchyards and cemeteries as in England, and nowhere exceeds 19 metres (about 60 feet) in height. The largest in girth, about 33 feet, at 3 feet from the ground, stands in the churchyard of Estry (Calvados). There are also two very fine trees at the church of La Lande Patny (Orne). Several others are figured by this author, of which the one at La Haye de Routot (Eure) is remarkable, on account of having in the interior of its hollow trunk a chapel about 6 feet in diameter and 10 feet high, which was built in 1806, and dedicated to Saint Anne des Ifs by the Bishop of Evreux.

In Germany, according to Willkomm,[6] the yew is most abundant in Pomerania, Hanover, and Thuringia, and he instances localities where it forms small pure woods. In the Darmbach forest district in the Eisenach Oberland there are, in

[1] Hansen, in *Jour. Roy. Hort. Soc.* xiv. 1892, p. 314.   [2] *Prodrome de la Flore Belge*, iii. 6 (1899).
[3] Mathieu, *Flore Forestière*, 509, 510 (1897).   [4] *Flora Pyrenæa*, I. 46 (1897).
[5] *Les vieux arbres de la Normandie*, iii. 359 (1895).   [6] *Forstliche Flora*, 275 (1897).

addition to many young plants, 311 yew trees of 1 foot or more in girth of stem. On the Veronica mountain at Angelroda in Thuringia, there are about 150 yew trees, of which the largest are possibly 600 years old. Apparently there were anciently two zones of distribution of the yew in Central Europe—a northern one which extended from the Netherlands through the coast provinces of Germany to the eastern shore of the Gulf of Riga, and a southern area comprising the mountainous regions of the Vosges, Jura, Black Forest, the whole of the Alps to Croatia, and the Carpathians. The yew also occurred in the hilly land of central Germany, where, at the present time, according to Drude,[1] it is indifferent to soil, as it grows on the muschelkalk near Göttingen, on the dolomite of Süntel in the Weser mountain district, and on primitive rock on the southern slope of the Rachel (up to 3300 feet altitude). On the dolomite it occurs as isolated trees, while in the ravines and rocky parts of Süntel it forms thick underwood. In the Bavarian Alps it ascends to 3800 feet, not being met with below 1240 feet.

In Switzerland the yew ascends in the Alps to 4660 feet. The largest and finest yew is at Geistler, near Burgdorf, at an elevation of 2400 feet above the sea. This tree is well figured in *Les Arbres de la Suisse*, t. xii., and is said to be 50 feet high by 12 feet in girth at 4 feet above the ground; it divides into several stems at about 10 feet up.

In Austria-Hungary the yew occurs in the Carpathians and the Alps, ascending in Transylvania to 5400 feet; and it is reported to occur in Roumania and Bulgaria.

The yew is generally distributed throughout the mountains of the Iberian Peninsula. In Spain, according to Laguna,[2] it almost always occurs as isolated trees, and is found in all the Cordilleras from the Sierra Nevada to the Pyrenees and the mountains of Asturias, also in the Balearic Isles. He has only seen it forming pure forest in the Sierra Mariola, near to Alcoy (Valencia). In the high part of that chain on its northern slope there exist what are called the *Teixeras de Agres*, groups of yews belonging to the town of Agres. Here, in 1870, there were still living some hundreds of ancient yews, with some young trees.

Gadow[3] says, "There are numerous large and small trees forming a scattered forest, between Riano and Cistierna at about 3600 feet elevation, the terrain belonging to the reddish Permian rubble. The yew tree is widely distributed throughout the Spanish mountains and on the Serra da Estrella (in Portugal), but is rare everywhere. Most of the trees are solitary and old, with decaying tops. Younger trees are ruthlessly destroyed by their branches being lopped off, to be used in the cattlefolds partly instead of straw, and partly for repairing the fences and roofs. The vernacular name is *Tejo*."

Willkomm[4] states that in the high mountains of Spain it occurs as isolated stunted trees, and says that on the Sierra de la Nieve there was an old yew tree which measured only 17 feet in height, although it had a girth of $17\frac{1}{2}$ feet. In the south of Spain it ascends to 6500 feet.

---

[1] *Hercynische Florenbezirk*, 114 (1902).   [2] *Flore Forestal Española*, i. 114 (1883).
[3] *Northern Spain*, 387 (1897).   [4] *Pflanzenverbreitung auf der Ibirischen Halbinzel*, 251 (1896).

I   P

In Italy, according to Piccioli,[1] the yew is found on the hills and in woods of the mountain regions of the Apennines and the Alps. It is only found in the maritime region in Liguria; but is common in Sardinia, where it ascends to 5660 feet. In Sicily it is found in the region of the olive, and occurs on Mount Etna, mixed with beech, to a height of 6000 feet. The yew, however, is not mentioned in Tornabene's *Flora Ætnea*.

In Greece[2] isolated trees occur in mountain woods up to the sub-alpine region. It is recorded from near Kastania, in Pindus, Mount Olympus, and Oeta (Thessaly); Mount Parnassus, Mount Malero (Laconia), and other places.

The yew[3] formerly occurred in the Azores, attaining timber size on Corvo and Flores, whence it was exported as a source of royal revenue. It is now apparently exterminated.

It occurs sporadically in the high mountains of Algeria,[4] in the Atlas of Blidah, Djurdjura, and Aures. A photograph of a venerable tree in Algeria, taken by M. de Vilmorin, is reproduced in *Garden and Forest*, 1896, p. 265.

In the Caucasus[5] it occurs throughout the whole territory, including Talysch, at altitudes varying from sea-level to 5660 feet.

In Asia Minor it occurs in Anatolia and Mysia, according to Boissier. Kotschy found it common in the Cilician Taurus from 6160 to 7600 feet altitude. Szovitch collected it in Armenia. It also occurs in North Persia.

Some wood[6] found in the palaces of Nineveh, and recorded on a tablet as having been brought as "cedar" from Lebanon, proved on microscopical examination to be yew.   (A. H.)

The yew, according to Gamble,[7] is a conspicuous tree in the Himalayan forests, at 6000 to 11,000 feet altitude from Afghanistan to Bhutan. It occurs in the Khasia Hills at 5000 feet, and in Upper Burmah at 5000 to 6000 feet. Sound trees are very scarce, but a very large one cut in Sikkim in 1876 was quite sound. Gamble has measured trees 20 feet in girth; one, 16 feet in girth, had a cylindrical bole 30 feet high. Madden records a tree at Gangutri, near the source of the Ganges, 100 feet high and 15 feet in girth, which surpasses anything I know of elsewhere. I have seen fine yews at 9000-10,000 feet on the Tonglo ridge, which divides Nepal and Sikkim, and have found many orchids upon them, one of which, *Cœlogyne ochracea*, has lived for 24 years in my collection. The growth in India varies from 23 to 55 rings per inch of radius. The timber weighs 46 to 59 lbs. per cubic foot, and is used for bows, carrying-poles, and native furniture, and if more common would be more extensively used, as it is very strong and elastic, and works and polishes beautifully. It requires, however, long seasoning.

Sir Joseph Hooker[8] noted that at 9500-10,000 feet on Tonglo the yew is an

[1] *Le Plante Legnose Italiane*, 31 (1890).   [2] Halácsy, *Consp. Fl. Græcæ*, iii. 459 (1904).
[3] Trelease, *Missouri Bot. Garden Ann. Report*, viii. 1897, p. 169.
[4] Battandier et Trabut, *Flore de l'Algérie*, 398 (1904).
[5] Radde, *Pflanzenverbreitung in den Kaukasusländern*, 183 (1899).
[6] G. Henslow in *Garden*, 1904, ii. 73.   [7] Gamble, *Indian Timbers*, 413.
[8] Hooker, *Himalayan Journals*, i. 168, 191, ii. 25 (1854).

immense tall tree with long sparse branches and slender drooping twigs, while at Choongtam (5000-6000 feet altitude) it is small and rigid, much resembling in appearance our churchyard yew. The red bark is used as a dye and for staining the foreheads of Brahmins in Nepaul.

There is a specimen at Kew, collected by Sir George Watt in Manipur, which bore yellow berries.

In the United States[1] there are a number of large European yew trees in New York, Philadelphia, and Baltimore, showing that the tree must have been brought to the eastern United States more than a century ago. Sargent says that everywhere south of Cape Cod it appears to be perfectly hardy. Farther east it suffers from the cold in severe winters, and cannot be considered a desirable tree for general planting in eastern New England. T. D. Hatfield,[2] writing from Wellesley in Massachusetts, states that the variegated form of the common yew is hardy in places where the green type perishes.

II. JAPANESE YEW, var. *cuspidata.*—*Ichii* in Japan, *Onko* of the Ainos in Hokkaido. Though Sargent says[3] that, judging from his observations, it is confined to the island of Yezo, it is stated in the *Forestry of Japan*, p. 88, that it is found also in Kiso and Nikko, and it was included in the list of trees growing wild in the royal forest of Kiso, though I did not see it myself. In Nikko it is planted in the temple gardens ; a fine specimen, of which I give an illustration taken at this place (Plate 53), shows how much it resembles our yew in habit and appearance. This tree was about 40 feet high by 12 in girth. In the Hokkaido it grows scattered through the lowland and hill forests, among deciduous trees and conifers, but nowhere, so far as I saw, gregariously, and attains a large size, trees of 50-60 feet high with clear trunks 2-3 feet in diameter being not very rare. It sometimes produces beautifully veined burrs, and when old is often rotten inside.

It is a favourite in gardens in Hokkaido, as trees of considerable size can be moved without killing them. The wood, which seems milder, sounder, and more free from holes and flaws than in England, is much used by the Japanese for water-tanks, pails, and baths, and is cut into handsome trays, sometimes carved, which I bought quite cheaply in Sapporo. I also procured large planks and slabs of it, measuring as much as 26 inches wide, and quite sound, such as I have never been able to get from English yews. Chopsticks, clogs, and the Aino bows are also made of yew wood, and when cut into thin shavings very pretty braid is made from it.

I was informed by Mr. N. Masaki of the Imperial Art School in Tokyo, that the semifossil wood known at Sendai as *Gindai-boku* is dug from the bed of the Natonigawa river, near which deposits of lignite are found. This wood was believed by the carvers at Nikko to be fossil Cryptomeria wood, but is so like the bog yew found in Great Britain in grain and colour that I have little doubt that it is yew. This wood is only procured in small pieces of irregular shape, the largest that I saw being made into a tray about 20 inches square. It is very hard, of

[1] *Garden and Forest*, 1897, p. 400. Large trees also occur at Washington, *loc. cit.* 1896, p. 261. [2] *Ibid.* p. 405.
[3] *Forest Flora of Japan*, p. 76.

a rich reddish brown colour, and when polished or carved is extremely hand-some.

The Japanese yew also occurs in Saghalien, the Kurile Isles,[1] Amurland, and Manchuria.   Apparently it is very variable in habit, as Maximowicz[2] regarded the Amurland plant as a mere shrub, though in one place in the mountains he saw a tree a foot in diameter.   Trautvetter[2] saw no difference between the yew in Amurland and in Europe, except that the seed of the former was smaller and more pointed.

The Japanese yew was introduced into England between the years 1854 and 1856 by Fortune,[3] who states that he received it from Mr. Beale in Shanghai, to whom it had been sent from Japan.   It was first cultivated and propagated by Mr. Glendinning of the Chiswick Nursery.   It has not grown to be a tree in England so far as we know, as it assumes rather the form of a large branching shrub with two or three stems.   It is usually distinguished from the other yews, as seen in cultivation, by the peculiar yellow colour of the under-surface of the leaves, which are broad, somewhat leathery in texture, and abruptly pointed.   This yellow colour is not, however, confined to the Japanese yew, as it occurs in the Chinese yew, and also apparently in some Pyrenean specimens, and is perhaps due to climatic influences.

According to Sargent[4] the Japanese yew was introduced into the eastern United States in 1862, and has proved to be perfectly hardy as far north as Boston. It grows rapidly in cultivation, and promises to become a large long-lived tree. Sargent speaks of a dwarf compact form of this plant with short dark green leaves in cultivation in the United States, which probably originated in Japanese gardens. It often appears under the name of *Taxus brevifolia*, but must not be confounded with the true *Taxus brevifolia* of the Pacific coast.   This is doubtless the *Taxus cuspidata*, var. *compacta*, of the Kew Hand List, of which we have seen no specimen. Sargent has also seen in California a yew with fastigiate, somewhat spreading branches, which had been imported from Japan, evidently another garden variety of *Taxus cuspidata*.

III. CHINESE YEW, var. *sinensis*.—The yew has only been found in China, in the provinces of Hupeh and Szechuan, where it is a very rare tree in the mountains at 6000 to 8000 feet, occurring on wooded cliffs.   The largest tree seen by Henry was about 20 feet in height, but with a girth of 7 or 8 feet.   The bark is almost a bright red in colour.   Franchet[5] considered the Chinese yew to resemble *Taxus cuspidata*, S. et Z., which in his opinion does not seem to differ from the European yew in any positive character.   The Chinese mountaineers reported the timber to be red, strong, and of fine quality, and called the tree *Kuan-yin-sha*, "the fir of the Goddess of Mercy."

IV. PACIFIC COAST YEW, var. *brevifolia*.—Though this tree was introduced by William Lobb in 1854,[6] it is still very rare, and we know no specimens of any size in

---

[1] Miyabe, "Flora of Kurile Isles," in *Mem. Boston Soc. Nat. Hist.* iv. 261 (1890).
[2] *Primitiæ Floræ Amurensis*, 259 (1859).
[3] *Gard. Chron.* 1860, p. 170.   Article by Fortune on Chinese Plants introduced during his travels in China in 1854-1856.
[4] *Garden and Forest*, 1897, p. 402.   [5] *Jour. de Bot.* 1899, p. 264.
[6] Veitch's *Man. Coniferæ*, ed. 1, 305 (1881).

England, though it might be so easily mistaken for the common yew, that we have possibly overlooked it. It would no doubt grow well in England, as it is a native of the colder and damper parts of the north-west coast of America, from Queen Charlotte Islands along the coast ranges of British Columbia, western Washington, and Oregon ; in California on the Sierra Nevada at 5000 to 8000 feet, and as far south as Monterey ; and extends eastward to the western slopes of the Rocky Mountains in Montana, where it becomes shrubby in habit. I have seen it in Washington on the slopes of Mount Tacoma, where it grew isolated in the dense forest, and attained no great size, though it occasionally reaches a height of 70 to 80 feet. In Vancouver's Island it is not uncommon in the rich, low meadows of the east coast, but the largest I saw were not over 30 to 40 feet high. The wood seemed indistinguishable from that of the European species, and was, like it, rotten at heart in old trees and full of holes. Sargent says that the Indians use it for bows, paddles, spear handles, and fish-hooks, but except for fencing posts it does not seem to be used by settlers.

V. CANADIAN YEW, var. *canadensis.*—This is only a creeping shrub with a stem occasionally a foot or two in height, and though it is said by Loudon to have been introduced in 1800, it has never obtained a place in English gardens. I have seen it common in Canada in thick forest, where it produced red berries very like those of our yew. Sargent gives its distribution as from Newfoundland to the northern shores of Lakes Superior and Winnipeg, southwards through the Northern States to New Jersey and Minnesota.

VI. FLORIDA YEW, var. *floridana.*—This is one of the rarest of North American trees, confined to a few localities in Western Florida, and, except by its habit, not easily distinguished from *T. canadensis.* It is usually shrubby, rarely attaining 25 feet high. It has never been introduced to cultivation in England, and is probably not hardy.

VII. MEXICAN YEW, var. *globosa.*—A tree about 20 feet in height, discovered in 1837 by Ehrenberg in Southern Mexico. There are also specimens at Kew of yews collected in Mexico by Hartweg, F. Müller, and W. Saunders, which vary considerably in foliage. This variety is scarcely known, as recent collectors have failed to rediscover the tree. It is very like the common yew.

(H. J. E.)

VARIETIES OF THE COMMON YEW IN CULTIVATION

These have in some cases originated as individual sports in the wild state ; in other cases they are due to the art of the gardener, who has greatly increased the number of varieties by selection. They differ from the type in various ways :—(1) in habit (fastigiate, prostrate, pendulous, and dwarf forms) ; (2) in the colour, shape, size, and disposition on the branchlets of the leaves ; (3) in the colour of the fruit. André,[1] in an interesting article, illustrated by figures, has drawn attention to the remarkable differences which occur in the shape of the seed and of the aril in the different culti- vated varieties ; but it is probable that these are not so constant as he believed.

[1] *Revue Horticole,* 1886, p. 105, translated in *Garden,* 1889, xxxv. 36.

A. *Fastigiate Forms.*—In these the branches take an upward direction (vertical or ascending), and the leaves tend to spread out radially from the branchlets.

1. Var. *fastigiata*, Irish Yew, Florencecourt Yew.

> *Taxus baccata fastigiata*, Loudon, *Arb. et Frut. Brit.* iv. 2066 (1838).
> *Taxus fastigiata*, Lindley, *Syn. Brit. Flora*, 241 (1829).
> *Taxus hibernica*, Hook. *ex* Loudon, *loc. cit.*

Columnar and compact in habit, all the branches and branchlets being directed vertically upwards. Branches stout, branchlets few and short. Leaves, always spreading radially in all directions around the branchlets, dark green and shining, with the apex usually more obtuse than in the common yew. Dr. Masters considers the Irish yew to be a juvenile form,[1] in which the characters of the seedling (the radial disposition of the leaves and the upright habit) are preserved throughout the life of the plant. As the original tree was a female, and the variety is propagated by cuttings, all Irish yews are of the same sex. When they bear flowers they are generally fertilised by the pollen of common yews growing in their neighbourhood, and the seed resulting, when planted, generally produces plants indistinguishable from the common yew.[2] Dr. Masters[1] received from Mr. Tillett of Sprowston, near Norwich, sprays of an Irish yew which bore male flowers. This was apparently an instance of a monœcious tree, a phenomenon which occurs though rarely in the common yew. No true male Irish yew has ever been met with. The aril of the Irish yew differs usually from that of the common form in being more oblong in shape.

The Irish yew was discovered[3] in the mountains of Fermanagh above Florencecourt by a farmer named Willis about the year 1780. He found two plants, one of which he planted in his own garden, and is now no longer living. The other was planted at Florencecourt, the seat of the Earl of Enniskillen; and from it cuttings were distributed, which are the source of all the Irish yews in cultivation. The original tree is still living, and a figure of it is given in Veitch's *Manual*, p. 141, as it appeared about thirty years ago. Kent says that in 1900 it had an open straggling appearance.

One of the finest Irish yews known to us is that at Seaforde, near Clough, Co. Down, the seat of Major W. G. Forde. This tree was reported to be 33 feet high in 1888,[4] and $35\frac{1}{2}$ feet in 1903.[5] A plate of it is given (Plate 58), reproduced from a photograph kindly sent us by the owner, who reports the present measurements (1905) to be :—Height, 37 feet; girth at the ground, 9 feet; circumference of branches at 20 feet from the ground, 91 feet.

Two large trees exist at Comber, Co. Down, of which Mr. Justice Andrews gives the following particulars in a letter :—

"The Irish upright yew trees at Comber, mentioned in Mackay's *Flora Hibernica* (1836), p. 260, are the two large yews[6] in the garden beside 'Araghmore,' the residence of Mrs. John Andrews. My earliest recollection of them goes back

---

[1] *Gard. Chron.* 1891, x. 68.

[2] Sir C. Strickland writes in *Gard. Chron.* 1877, vii. 151 : "All the plants I have raised from Irish yew berries are exactly like the common yew." But Elwes saw at Orton Hall three seedlings from the Irish yew of which one was fastigiate in habit.   [3] *Gard. Chron.* 1873, p. 1336.   [4] *Ibid.* 1888, iv. 484.

[5] *Ibid.* 1903, xxxiii. 60.   [6] Loudon figures one of these on p. 2067.

60 or 70 years, and they were then apparently as tall as they are now, but not so much spread out. I cannot accurately estimate their height and girth, but they are the two largest upright yews I have seen."

At Brockhill,[1] Worcester, there are two large Irish yews, estimated by Mr. Lees to have been at least 100 years old. Very handsome specimens are also growing at Montacute House, Somerset.

The Irish yew is very effective as a garden tree, but requires pruning and wiring every two or three years in order to keep it in good shape. There is at Colesborne a terrace planted on both sides with Irish yews about 50 years ago, which are kept in shape by wire, and when so treated are of very uniform growth and habit.

Taxus fastigiata aurea is a form of the Irish yew, in which the young shoots are golden yellow. In *Taxus fastigiata argentea* the tips of the branchlets are white.

2. Var. *Chesthuntensis*.

Taxus baccata Chesthuntensis, Gordon, *Pinet. Suppl.* 98 (1862).

This was raised by William Paul of the Cheshunt Nursery from a seed of the Irish yew, which it resembles. The branches, however, are ascending, but not so erect as in the parent form. The leaves have an acute apex, and resemble in colour those of the Irish yew, being dark green and shining on the upper surface; they are broader and shorter than those of the common yew. It is less formal than the Irish yew, and is said to grow twice as fast.

3. Var. *elegantissima*.

—This was raised, according to Barron,[2] by Fox of the Wetley Rock Nurseries, who had an Irish and a golden yew growing together, from which this came as a seedling. It is generally a dense compact shrub, but forms occur which are more open in habit. The leaves are usually radially spreading, but are often two-ranked on some of the branchlets; they are long, and terminate gradually in a long, fine cartilaginous point. Young leaves are golden yellow; adult leaves have white margins.

4. Var. *erecta*.

Taxus baccata erecta, Loudon, *loc. cit.* 2068 (1838).
Taxus baccata Crowderi, Gordon, *Pin. Suppl.* 98 (1862).
Taxus baccata stricta, Hort.

A dense broad shrub with erect and ascending branches. The leaves are dark green, shining, short, and acute; and are usually radially arranged, but often on the lower branchlets are disposed in two ranks.

The Nidpath Yew[3] resembles this variety, but is more columnar in habit, with a tendency to spread at the top. The leaves, as seen on a shrub at Kew, are bluish green, and usually are all radially arranged.

A variety named *imperialis* is described as being a slender, tall form with ascending branches and dark green leaves.

[1] *Trans. Worcester Nat. Hist. Club*, 1847-1896, p. 211.
[2] *Gard. Chron.* 1868, p. 921. Veitch's *Manual*, 1st ed. 302, states that it was introduced by Messrs. Fisher, Son, and Sibray of the Handsworth Nurseries, near Sheffield. [3] Nicholson, *Dict. of Gardening*, iv. 12.

B. *Dwarf forms with leaves radially arranged on the branchlets.*

5. Var. *ericoides*.

> *Taxus baccata ericoides*, Carrière, *Conif.* 519 (1855).
> *Taxus baccata empetrifolia*, Hort.

A low shrub with ascending branches. Leaves generally radially arranged, but occasionally two-ranked, uniform in size, falcate, short, acute, tapering to a fine cartilaginous point.

6. Var. *nana*.

> *Taxus baccata nana*, Knight, *Syn. Conif.* 52.
> *Taxus Foxii*, Hort.

A dwarf shrub. Leaves generally radially arranged, some being two-ranked; very variable in length, but always short, straight or falcate, often twisted or curved.

C. *Varieties with leaves distichously arranged, assuming pendulous, prostrate, and other non-fastigiate habits.*

7. Var. *Dovastoni*, Dovaston Yew.

> *Taxus baccata Dovastoni*, Loudon, *loc. cit.*, 2082 (1838).

A tree or large shrub, with spreading branches, arising in verticils, and becoming very pendulous at their extremities. Leaves dark green with an abruptly mucronate apex.

An account of the original tree, from which this variety has been propagated, is given in Loudon and in Leighton's *Flora of Shropshire*.[1] This tree was planted as a seedling about the year 1777 at Westfelton, near Shrewsbury. It was in vigorous health in 1900, and measured then 8 feet 10 inches in girth at $4\frac{1}{2}$ feet from the ground. Nineteen years previously its girth was 7 feet 11 inches. It is described as having a single leader, with branches pendulous to the ground. The original tree is monœcious; one branch only producing fertile berries, from which seedlings were raised, which reproduced the habit of the parent.[1] Barron[2] states that all his Dovaston yews are female trees. Carrière[3] sowed seeds of this form on many occasions, and the offspring was always like the common yew, doubtless due to his Dovaston yews being fertilised by the pollen of ordinary yew trees in the vicinity.

Carrière further states that MM. Thibaut and Keteleer obtained in 1865, from seeds of this variety, plants which were in the proportion of three-fourths variegated in foliage and one-fourth green; but in no case was the pendulous habit observed. The variegated plants passed into commerce as *Dovastoni variegata*; but these were simply ordinary variegated yews. A sub-variety, however, occurs in which the leaves of the Dovaston yew are variegated with yellow; and this is known as var. *Dovastoni aureo-variegata*.

8. Var. *pendula*.—Growing at Kew, this is an irregularly branching wide, low,

---

[1] *Gard. Chron.* 1900, xxvii. p. 146, where a figure and full details of the Dovaston yew are given.

[2] *Ibid.* 1868, p. 992. He gives the dimensions of the Westfelton tree in 1876 as 34 feet high by $7\frac{1}{2}$ feet in girth. *Garden*, ix. 341.

[3] *Traité gén. des Conifères*, ii. 763 (1867).

dense shrub, making no definite leader, with the tips of the branchlets pendulous. Var. *gracilis pendula* is said to have the branches and branchlets more elongated, and to attain a larger size than var. *pendula*.

9. Var. *horizontalis*.

Taxus baccata horizontalis, Knight, *Syn. Conif.* 52 (1850).

This resembles the Dovaston yew in the verticillate arrangement of the spreading branches. The branchlets, however, instead of being pendulous, are turned slightly upwards at the ends of the branches.

10. Var. *recurvata*.

Taxus baccata recurvata, Carrière, *Conif.* 520 (1855).

A large shrub, with branches somewhat ascending and elongated, and pendulous branchlets, which bear the leaves so arranged as to be all directed upwards, each leaf being recurved. The leaves resemble those of the Dovaston yew.

11. Var. *procumbens*.

Taxus baccata procumbens, Loudon, *loc. cit.* 2067 (1838).

A low prostrate shrub, keeping close to the ground, with branches long and ramified. This is distinct from *Taxus canadensis* in characters of leaves and buds.

D. *Varieties with leaves distichously arranged, in which the leaves are variously coloured.*

12. Var. *aurea*, Golden Yew.

Taxus baccata aurea, Carrière, *Conif.* 518 (1855).

A golden yew is mentioned in Plot's *History of Staffordshire* as occurring in that county in 1686. There are many kinds of golden yew, which are of different origin. The form generally known as *aurea* is a dense shrub or low tree, with narrow falcate leaves which are variegated with yellow. Golden yews of this kind are said to be all male trees. The original was reared by Lee of Hammersmith, and was afterwards planted at Elvaston Castle. It was monoecious,[1] and from it Barron reared several varieties. The variety known as var. *Barroni* has the leaves more decidedly yellow than those of the common golden yew; and one form of it is female and bears berries.

A great number of variegated yews of different kinds have been raised at Knap Hill, at the Chester Nurseries, and elsewhere. These have been obtained as seedlings in various ways; some were obtained by planting Irish yew amongst common golden yew; in other cases the seed-plants used were varieties like *elegantissima*, *erecta*, *adpressa*, etc.

13. Var. *Washingtoni*.—A low dense shrub, in which the leaves on the young shoots are golden yellow in colour.

---

[1] According to Barron the tree was a male; but he discovered on it a single branch bearing female flowers. See *Gard. Chron.* 1868, p. 921; also 1882, ii. 238.

I

14. Var. *glauca*.

> *Taxus baccata glauca*, Carrière, *Conif.* 519 (1855).

A vigorous shrub, with leaves, which are shining and dark green on the upper surface, and glaucous blue beneath.

E. *Variety with differently coloured fruit.*

15. Var. *fructu luteo*.

> *Taxus baccata fructu luteo*; Loudon, *loc. cit.* iv. 2068 (1838).

This variety only differs from the common yew in the aril of the fruit being yellow. A tree of this kind was discovered about the year 1817 at Glasnevin, near Dublin, growing on the property of the Bishop of Kildare.

Cuttings, however, were first distributed from a tree noticed in the grounds of Clontarf Castle in 1838. This tree[1] was about 50 feet high in 1888. At Ardsallagh, Co. Meath, the residence of Mrs. M'Cann, there is a tree 30 feet high and 7 feet in girth, with yellow fruit, occurring in an avenue of old yews. There are several trees of this kind at Powerscourt,[2] the best one of which was about 40 feet high in 1888. Bushes raised from the seeds of these trees are reported to be bearing yellow berries, from which it would appear that this variety comes true from seed. It is remarkable that all the yellow-berried yews known, except the one mentioned above as collected at Manipur, should occur in the neighbourhood of Dublin.

F. *Variety with small leaves.*

16. Var. *adpressa*.

> *Taxus baccata adpressa*, Carrière, *Rev. Horticole*, 1855, p. 93; *Taxus adpressa*, Gordon, *Pinetum*, 310.
> *Taxus tardiva*, Lawson, *ex* Henkel and Hochstetter, *Syn. Nadelh.* 361.
> *Taxus sinensis tardiva*, Knight, *Syn. Conif.* 52 (1850).

A large spreading shrub with densely crowded branchlets, bearing remarkably small broad leaves, arranged on the shoots, as in the common yew. The leaves are dark green above, $\frac{1}{4}-\frac{1}{2}$ inch long, elliptic linear in outline, with a rounded apex, from which is given off a short mucro. The aril is broad and shallow, not covering the seed, which is 3-angled and often depressed at the summit.

This is by far the most distinct of all the forms, geographical and horticultural, not only in foliage, but also in fruit. It has been considered by many botanists to be a distinct species, conjecturally of Japanese or Chinese origin. It is not known in Japan,[3] except as a plant introduced from Europe; and there is no reason for doubting the positive information[4] as to its origin given by Messrs. James Dickson and Sons and by the late Mr. F. T. Dickson of Chester, though there is a slight discrepancy in their two accounts. The latter states that it was found as a seedling by his father amidst some yew seedlings about 1838, while the former give

---

[1] *Gard. Chron.* 1888, iv. 576.
[2] *Ibid.* 707.
[3] Matsumura, *Shokubutsu Mei-I.* 290 (1895).
[4] *Gard. Chron.* 1886, xxix. 221, 268.

the date as 1828, and the locality as a bed of thorn seedlings in the Bache Nurseries, Chester.

Only female plants of this variety are known, and it is reproduced by grafting. Its flowers are doubtless fertilised by the pollen of common yew trees near at hand, and as a rule it produces a great crop of berries. Messrs. Dickson and Sons have frequently sown seeds which invariably produced the common yew.

Var. *adpressa stricta* is a form of this variety in which the branches are erect or ascending. It is not known whether it originated as a seedling or as a sport fixed by grafting. It was raised by Mr. Standish.

Var. *adpressa aurea* is a form with golden leaves.

Var. *adpressa variegata* is a form with the young shoots suffused with a silvery yellow colour. This was exhibited at the Royal Horticultural Society on August 27, 1889.

There are fine examples of var. *adpressa* in Kew Gardens.

SEEDLING [1]

The two cotyledons, together with the seed-case which envelops them as a cap, are carried above ground by the lengthening caulicle; and speedily casting off the remains of the seed-case, act as if they were true leaves. They differ from the latter in bearing stomata on the upper and not on the lower surface, and in having their apices rounded and not acute. The young stem, angled by the decurrent bases of the leaves, gives off at first three or four opposite pairs of true leaves, which are succeeded in vigorous plants by a few alternate leaves, crowded at the summit around a terminal bud, which in all cases closes the first season's growth, when the young plant is 1 to 3 inches high. The caulicle, 1 to 2 inches in length, ends in a strong tap-root, which descends several inches into the soil, and gives off a few lateral fibres.

The growth of the seedling during the next four or five years is very slow, often scarcely an inch annually. Afterwards the growth becomes more rapid.

SEXES, FLOWERS, FRUIT, BUDS

The yew is normally diœcious; but exceptions occur, and in our account of the cultivated varieties two or three instances of monœcious trees have been mentioned. The celebrated yew at Buckland,[2] Kent, is monœcious. As a rule it is only a single twig or branch which bears flowers of a different sex from those on the rest of the tree. A yew[3] at Hohenheimer, near Stuttgart, is reported, however, to bear male and female flowers irregularly over the whole tree, each kind, however, on separate twigs. There is a specimen at Kew of a branch, sent in 1885 by the Rev. T. J. C. Valpy of Elsing, Norfolk, which bears both male flowers and fruit.

---

[1] Figured in Lubbock, *Seedlings*, ii. 553, fig. 677 (1892).
[2] *Gard. Chron.* 1880, xiii. 556. There are specimens of this yew in the Kew herbarium.
[3] Kirchner, *loc. cit.* 74.

Gilbert White thought that male trees are more robust in growth than female trees ; but we are unaware of any accurate observations on this subject. Kirchner,[1] however, states that there is a slight distinction in the habit of the two sexes, male trees being taller with longer internodes and shorter leaves.

In early spring drops of mucilage may be observed glistening on the ovules of female trees in flower. The mucilage is secreted by the micropyle, and seems to entangle the grains of pollen which have been wafted on the ovules by the wind. The clouds of pollen which fly forth from the male flowers are well known. The pollen is liberated from the stamens by a very elaborate mechanism, which serves to protect the pollen grains in rainy weather. A good account of this is given by Kerner.[2]

A large quantity of fruit of the yew falls to the ground in autumn ; but the seeds in this case do not as a rule germinate. Natural reproduction seems to be effected by birds like the thrush and blackbird, which, attracted by the fleshy aril, devour the whole fruit. The seeds, protected by their hard testa, escape digestion and are voided at a distance. They rarely germinate in the first year after ripening ; seedlings come up as a rule in the year following, a few even appearing in the third year.

The buds of the different geographical forms appear to differ more than the leaves themselves. The terminal bud is invested closely by the uppermost and youngest leaves and continues the growth of the shoot. The bud scales on unfolding remain at the base of the growing shoot, and on older branchlets persist as dry brownish scales, forming an involucre at the bases of the branchlets. Lateral buds are developed on the twigs at irregular intervals. Many of these remain dormant, retaining the power to take on growth at any moment. This explains the readiness with which the yew submits to pruning, and the facility with which it produces coppice shoots when the stem is cut. Spray or epicormic branches are frequently produced on the upper side of the branches or on the stem ; and these also originate in dormant buds.

True root-suckers are never formed ; but layering occurs, though very rarely, in branches which have come in contact with the ground.   (A. H.)

Age, Hardiness

With regard to the supposed great age of yew trees, which has been much exaggerated by authors—especially by the great Swiss botanist, De Candolle—we must refer our readers to Lowe, who has discussed the subject very thoroughly in chapter iii. of his work. He proves that the average rate of growth is about 1 foot of diameter in 60-70 years in both young and old trees. There is, however, abundant evidence to show that though old trees grow at intervals much more rapidly than young ones, they do not grow uniformly, but have periods of comparative rest, and that the increase of girth is fastest when old trees have lost their heads and the stem is covered with young shoots.

[1] Kirchner, *loc. cit.* 74.   [2] *Nat. Hist. Plants*, Eng. translation, ii. 145, 146 (1898).

No tree has such a remarkable faculty of covering up wounds or injuries by the growth of fresh wood from the outside ; and even after the main stem is completely dead, fresh and entirely new stems may grow up around it and form a new tree around the dead one. For this reason most of the yews of very large size are mere shells, and even when no hollow can be seen from the outside, decay—which is often indicated by moisture running from holes in the trunk—has set in.

Three very curious sections showing the way in which these trunks grow are given by Lowe, pp. 78 and 79.

The yew, though occurring wild far north, as in Norway, is not perfectly hardy, and many instances are on record in which it has been injured or killed during severe winters. It was affected in Cambridgeshire[1] and severely injured at Glasgow by the severe frost of 1837-1838. In the winter of 1859-1860 the young shoots of many trees were killed at Burton-on-Trent.[2]

Many cultivated yews[3] were killed by the frost of 1879-1880 in Switzerland, Rhineland, Hessia, Thuringia, etc. though in the same localities other native conifers were not injured by the severe cold. Duhamel[4] states that in France the yew suffered much damage from the great frosts of 1709 ; and Malesherbes found several killed by the frost of 1789.

POISONOUS PROPERTIES OF THE YEW

The poisonous properties of the yew have been well known from the earliest times, and the subject has been so carefully investigated in the *Journal of the Royal Agricultural Society of England*, 1892, p. 698, by Messrs. E. P. Squarey, Charles Whitehead, W. Carruthers, F.R.S., and Dr. Munro, and summarised by Low in chapter x. of his work, that we need not do more than give a brief résumé of the present state of our knowledge. Through the kindness of Sir W. Thiselton Dyer, we have been able to peruse a file of the Board of Agriculture entitled " Yew Poisoning," in which the subject has been further discussed by that gentleman with whose opinions we are in complete accordance.

The conclusions drawn by Dr. Munro, after careful study from a medical point of view, are as follows :—

" Both male and female yew leaves contain an alkaloid.

" This alkaloid in both cases appears to agree with the taxine of Hilger and Brande. Taxine is probably the poison of the yew, but it is doubtful whether it has ever been obtained in a pure state, and its physiological effects have not been sufficiently studied. Other alkaloids are probably present in yew.

" Taxine is present in fresh yew leaves as well as in those withered or air-dried. It is also present in the seeds, but not in the fleshy part of the fruit.

" The yew poison may be one of moderate virulence only, and may occur in greater percentage in male than in female trees, or the percentage may vary from tree to tree without distinction of sex, and this may explain the capricious

---

[1] Lindley, *Trans. Hort. Soc.* 1842, ii. 225.   [2] *Gard. Chron.* 1860, p. 578.
[3] Kirchner, *loc. cit.* 62.   [4] *Traité de Arbres*, i. 302 (1755).

occurrence of poisoning. Also the half-dried leaves would be, *cæteris paribus*, more potent than the fresh.

"Further and extended chemical researches, in conjunction with physiological experiments, are necessary to clear up the matter.

"The principle having a specific uterine action is possibly not the same as that which causes death."

This poison, if taken in sufficient quantity, is deadly to man, horses, asses, sheep, cattle, pigs, pheasants, and possibly other animals, but under ordinary circumstances small quantities of the leaves may be and are habitually eaten by live-stock without apparent injury, whilst it seems proved that the wood of the yew may be used for water vessels and for baths, as in Japan, without any deleterious effects.

Sargent, *Silva of North America*, vol. x., 63, says "no cases of poisoning by *Taxus* in North America appear to be recorded"; and Brandis, *Forest Flora of British India*, p. 541, says that "in India domestic animals are said to browse upon *T. baccata* without experiencing any bad effects."

With regard to the danger of allowing this tree to grow in hedges and fields where stock are pastured, there seems to be abundant evidence, which is well summarised by Mr. E. P. Squarey, and which my own experience entirely confirms, that though animals which have been bred and fed in places where they have access to yew are more or less immune, probably because they never eat it in sufficient quantity to do harm, yet that animals freshly turned into such places when hungry, or in winter and spring when there is little grass, are liable to die from eating it, and that fatal effects most commonly ensue when loppings or partially withered branches and leaves are eaten.

It has been held in more than one case that landowners, and others responsible for keeping up fences, who allow yew trees to remain insufficiently fenced, are liable to an action for damages if another person's cattle from adjoining land eat the branches and die.

With regard to the danger of yew trees in game coverts we have little exact knowledge, but in certain cases there seems to be evidence of its being poisonous to pheasants, and the following passage, which was communicated by Sir William ffolkes, Bart., of Hillington Hall, Norfolk, to the editor of the *Journal of the Royal Agricultural Society*, 1892, p. 698, is worth quoting in full.

"Some years ago, when shooting through the coverts here the second time, we found about fifteen carcasses of pheasants under some yew trees. These could not have been overlooked the first time in picking up, as there was no stand anywhere near this place where so many pheasants could have been shot. My keeper informs me that it is after the pheasants have been disturbed by shooting that they take to perching in the yew trees. This may or may not be so, but at any rate it appears that, when they take to perching in these trees, they are apt to eat a few of the leaves. We now always drive them off the yew trees when they go to perch at night. I enclose some of the yew which poisoned the pheasants, and would like to add that never before this year have we picked up a dead pheasant anywhere near these yew trees till the coverts had been shot."

We have no record of any case of deer being poisoned by yew, though no doubt in a heavy snow they might be tempted to eat it, and Mr. Squarey states that in the "Great and Little Yews" of which I write later, hares and rabbits, which are very numerous, have never been found dead from poisoning.

I may add that I have frequently seen yews of a few inches in girth barked and killed by rabbits where they are very numerous and hungry, but it is one of the last trees to be attacked.

CULTIVATION

The yew is best raised from seed, except in the case of varieties which are propagated by cuttings, which are taken off in April or August and put into sandy soil in a shady border, or, better, under a handlight, as they will then root more quickly.

Seed, if sown when ripe, will sometimes come up in the following spring, but usually lies over the first year, and is therefore treated like haws. The seedlings grow very slowly at first, and require several years of nursery cultivation before they are large enough to plant out.

They are easy to transplant in early autumn or in spring, and may be safely moved at almost any time of the year even when of large size, if care is taken to prepare the roots and keep them watered until new ones are formed.

The yew in Buckland Churchyard, about a mile from Dover, may be mentioned as an instance of the great age at which this tree may be transplanted with safety, if proper care and appliances are used. This tree was a very old and large one, divided into two stems, one of which, almost horizontal, was 10 feet 10 inches, and the entire trunk no less than 22 feet in girth. It was removed by the late Mr. W. Barron on March 1, 1880, to a position 60 yards off, where Mr. John Barron of Elvaston Nurseries tells me it is now in a vigorous state of health. An account of this tree is given by Lowe; and the manner in which it was transplanted, with pictures of its appearance before and after removal, is described fully in *Gard. Chron.* 1880, p. 556-7.

By sowing seeds there is some chance of obtaining variegated forms, which are among the most ornamental shrubs we have.

The Hon. Vicary Gibbs has found that at Aldenham the use of nitrate of soda increases in a marked degree the growth of young yew trees. Some yews planted by him in 1897 and treated with liberal quantities of this manure had attained in 1905 an average height of 12 feet, with a girth of stem of 16 inches at a foot above the base.

SOIL AND SITUATION

Though the yew grows naturally most commonly on limestone formations in England, it will grow on almost any soil except perhaps pure peat and wet clay, and attains its largest dimensions on deep sandy loam. It grows better under dense shade than any tree we have, and may therefore be used for underplanting beech-woods where bare ground is objected to, and where the soil is too poor and dry or too limy for silver fir. In such situations, however, it grows very slowly and produces little or no fruit.

REMARKABLE TREES

No tree, except perhaps the oak, has a larger literature in English than the yew ; and though a monograph on the *Yew Trees of Great Britain and Ireland* by the late John Lowe, M.D., was published by Macmillan so lately as 1897, I am able to add many records of trees not known to him, and shall not allude to most of the trees which he has described and figured.

It is strange that neither Loudon, Lowe, nor any other writer has, so far as I know, described the yews in the close walks at Midhurst, which, on account of their extraordinary height, form what I believe to be the most remarkable yew-grove in Great Britain or elsewhere.

The age and history of these wonderful trees is lost in obscurity, but it is said in Wm. Roundell's very interesting book on Cowdray[1] that Queen Elizabeth was entertained at a banquet in these walks, so they must have been of considerable age and size 300 years ago.

The close walks are situated close to the town on the other side of the river, and consist of four avenues of yew trees forming a square of about 150 yards, together with a grove of yews at the upper end which average, as nearly as I could measure them, about 75 feet in height, but some probably exceed 80. These trees are for the most part sound and healthy, though little care has been taken of them, and some have fallen. They are remarkable not only for their great height, which exceeds that of any other yews on record in Europe, but on account of their freedom from large branches, many having clean boles of 20-30 feet with a girth of 8-9 feet. They stand so thick together that on an area of about half an acre or less—I made 213 paces in going round it—I counted about 100 trees and saw the stumps of 10 or 12 more, which would probably average over 30 cubic feet to each tree without reckoning the branches.

The ground below is absolutely bare of vegetation, and though I found some small seedlings among the grass and briars on the outside of this area, I do not think the yew grows from seed under its own shade.

The photographs (Plates 54, 55) will give a fair idea of the appearance of this wonderful grove, and of the walks which lead to it. Some of the trees have a remarkable spiral twist in them like fluted columns, which I have not seen so well developed elsewhere.

The soil on which they stand seems to be of a light sandy nature, but deep enough to grow large fine timber of other species, and is, I believe, on the Lower Greensand formation.

[1] Cf. *Guide to Midhurst*, p. 41 (Midhurst : G. Roynon (1903)).

Another and perhaps the largest pure yew-wood in England is on the downs three miles west of Downton, Wilts, on the property of the Earl of Radnor. It is known as "The Great Yews," and contains about 80 acres. The trees are not remarkable for their size, and appear to have been partly planted, as the largest are at regular intervals and of about the same age. Probably at a time when yew - wood was wanted for bow - making an existing wood was filled up with planted trees, and no doubt these yews could tell some striking tales. Mr. E. P. Squarey, who took me to see them, and who has seen little change in them during the last 60 years, pointed out one under which some tramps had been caught in the act of roasting a sheep they had stolen, and related various tragedies which had occurred in this wild district in bygone times.

"The Little Yews" is the name of another wood about half a mile from the "Great Yews," which, though not of such large extent, contains much finer trees, many being from 8 feet to 10 feet in girth and 50 feet high. As in other yew woods (at any rate where rabbits exist) I found few or no young trees coming up, and the mixture of beech, ash, oak, thorn, whitebeam, and holly trees which are found in the more open spaces all appear to be self-sown. Several of the largest trees have been recently blown down.

After the Midhurst and the Great and Little Yews, I think the Cherkley Court Yew Wood is the best in England; and, thanks to the kindness of A. Dixon, Esq., the owner, I am able to give some particulars of this interesting place, which Lowe thought to be the finest collection of yews in existence.

The wood covers an area of 50 to 60 acres in a shallow valley forming part of the old Ashurst estate, about three miles from Leatherhead in Surrey, on the east side of the old pilgrims' road to Canterbury. It was formerly a rabbit warren, but is now carefully preserved by Mr. Dixon. It is said that 500 yew trees were once sold out of this wood by Mr. Boxall for 10 guineas each, and these two facts will probably account for the fact that there are now scarcely any young trees coming up, and but few trees with straight, tall trunks. Their average height does not exceed about 40 feet, and the majority of them are not well-grown trees, but there are some of great girth, of which the best is called the Queen Yew, and measures 14 feet 6 inches at 1 foot from the ground; then swelling out in a peculiar way and measuring 20 feet 4 inches at about 4 feet. At this height it begins to branch, and though the main stem goes up some way, the whole tree is certainly under 50 feet in height.

One of the most curious trees in this grove, called the Cauliflower Yew, was figured in the *Gardeners' Chronicle*, and copied in Veitch's *Coniferæ*, ed. ii. p. 128. This tree has now lost much of its beauty, owing to a heavy snowstorm which occurred in 1884 and which did serious damage to the Cherkley Yew Wood.

Another place of great interest to naturalists, where the yew is in great abundance, is Castle Eden Dene, in Durham, the property of Rowland Burdon, Esq. This locality is renowned among botanists as the last in England where the ladies' slipper orchid (*Cypripedium Calceolus*) still exists. It is a deep valley about 3 to 4 miles long, running down to the sea, and, in some places, has steep cliffs of a peculiar magnesian limestone formation, which decomposes into a red loamy soil, on which yews grow very freely, though they do not attain anything like the size that they do in the south of England. The largest which I measured was only 8 feet in girth. What makes them so picturesque is the way in which their roots spread over the bare rocks, and the mixture of curiously gnarled wych elms which accompany them. All the foot-bridges are here made of yew wood, but it is not cut except for home use, and is not increasing by seed—again, I think, on account of the rabbits.

There is a remarkably fine yew walk at Hatherop Castle, Gloucestershire, the seat of G. Bazley, Esq., which is supposed to be about 300 years old, in which the trees average about 60 feet in height with a girth of 9 to 12 feet.

The yew in Harlington Churchyard, near Hounslow, Middlesex, was considered by Kirchner (*loc. cit.* 60) to be the tallest yew tree in Europe, viz., 17.4 metres (57 feet). Lowe, page 85, gives the height in 1895 as 80 feet, on the authority of the Rev. E. J. Haddon. Henry saw this yew in October 1895, and measured the height as 50 feet only, and this is correct, within a margin of error of less than 2 feet. This tree is 17 feet 3 inches in girth at the base, where the bole is narrowest; above this it swells and is very gnarled, and at 10 feet up it divides into two great limbs.

A celebrated yew stands in the churchyard at Crowhurst, in Surrey, and has been described by Lowe (p. 201) and figured by Clayton.[1]

Crowhurst, in Sussex, has another great old tree of which much has been written, and which Low figures (p. 38).

One of the finest yews in England is the Darley yew, growing in the church-yard at Darley Dale, Derbyshire. From a work on Derbyshire Churches, by the Rev. J. C. Cox, M.A., which has been sent me by Messrs. Smith, the well-known nurserymen of Darley Dale, I abridge the following particulars of it :—The churchyard is celebrated for what is claimed to be the finest existing yew tree in England, or even in the United Kingdom. Rhodes, writing of it in 1817, says that the trunk for about 4 yards from the ground measures upwards of 34 feet in girth; but Lowe gives (p. 207) measurements taken by four different persons between 1836 and 1888, of which the largest is 34 feet 6 inches by Mr. Smith in 1879, and the most recent and exact by Mr. Paget Bowman in 1888, which gives 32 feet 3 inches at 4 feet from the ground. This gentleman cut from it with a trephine nine cylinders of wood on one horizontal line which show 33 to 66 rings per inch of radius, showing an average growth of an inch in 46 years. There is a cavity in the tree about half-way up one of the trunks which will hold seven or eight men standing upright. At the ground the girth is 27 feet, and at this point no increase has taken place for 52 years. The height is not given, but a photograph by Mr. Statham shows it as about 50 feet.

I have chosen the tree at Tisbury for illustration as a specimen of the church-yard yew, for though figured by Lowe, his plate gives a poor idea of its symmetry, and it is one of the largest healthy yews in England. Though difficult to measure on account of the young spray which its trunk throws out, I made it in 1903 to be

[1] *Trans. Bot. Soc. Edinburgh,* 1903, p. 408.

about 45 feet high by 35 in girth. The trunk is hollow, and has inside it a good-sized younger stem, probably formed by a root descending inside the hollow trunk from one of the limbs. It is a female tree, and of its age it is impossible to form a correct estimate. (Plate 35.)

At Kyre Park, Worcestershire, the residence of Mrs. Baldwyn Childe, is a very fine yew tree growing near the wonderful grove of oaks which I have described elsewhere; it measures 55 feet high by 20 in girth. Under it the Court Leet of the Manor was formerly held.

The most widespreading yew I have seen is a tree at Whittinghame, the seat of Mr. Arthur Balfour, which I measured in February 1905. It grows near the old tower, formerly the property of Sir Archibald Douglas, one of the conspirators of Darnley's murder, and, according to a local tradition, this was plotted under its shade. The tree is not remarkable for height or girth, the bole being only about 12 feet high and 10½ feet in girth, but spreads out into an immense drooping head, the branches descending to the ground and forming a complete circular cage or bower about 10 yards in diameter, inside which, Mr. Garrett, the gardener, told me that 300 school children had stood at once. The branches lie on the ground without rooting, so far as I could see, and spread so widely that I made the total circumference about 110 paces. Mr. Garrett, with a tape, made it 125 yards. The appearance of the tree from outside is fairly well shown in Plate 36.

Another tree of this character, but not so large, grows at Crom Castle, on upper Lough Erne, and is described in the *Ulster Journal of Archæology* by Lord Erne.[1] It is said to resemble an enormous green mushroom in contour, and has evidently been a trained tree, its horizontal branches being supported on timber supports upheld by about 60 stout props. Its total height is given as 25 feet, with a bole of 6 feet and a girth of 12 feet, the branches being 250 feet in circumference.

Yew trees in a wild state do not, as a rule, grow so large as those which are planted, probably because they are usually in poor rocky soil and crowded by other trees; but Lord Moreton tells me of a remarkably fine one which was shown him by Mr. Roderick Mackenzie, son of the owner of Fawley Court, in a wood on the Greenlands property on the Chiltern Hills. He described the tree as of the most symmetrical growth, and he guessed it to be nearer 70 feet than 60 feet high, with a girth of about 12 feet.

YEW HEDGES AND TOPIARY WORK

The yew, owing to the readiness with which it submits to pruning, forms an admirable hedge, and an excellent account of the conditions necessary to success in the making and keeping up of yew hedges is given by Mr. J. Clark in recent issues of the *Garden*,[2] to which we refer our readers.

One of the oldest and finest yew hedges in Great Britain is that at Wrest Park,[3]

---

[1] *Cf.* Loudon, *loc. cit.* 2081.   [2] *Garden,* 1905, lxvii. 54 and 136.   [3] *Gard. Chron.* 1900, xxvii. 375.

which is said to be 350 years old. There is a very high one of semicircular form enclosing the approach to the front door of Earl Bathurst's house at Cirencester.

Others[1] occur at Pewsey in Wiltshire, Melbourne in Derbyshire, Holme Lacy near Hereford, Hadham in Hertford, Albury Park near Guildford, etc.

An interesting account of the use of yew in topiary work is given by Kent,[1] who gives two illustrations of the remarkable effects produced by this art at Elvaston Castle. Leven's Hall,[2] Westmoreland, is also noted for the extraordinary forms into which the yew has been forced to grow. In a recent work[3] by Elgood and Jekyll pictures are given of several remarkable effects produced by the yew, notably the Yew Alley at Rockingham, the Yew Walk at Crathes, and the Yew Arbour at Lyde.

There are some remarkable clipped yews in the garden at Gwydyr Castle, in the valley of the Conway, a beautiful old place now belonging to Lord Carrington, which have been the subject of careful attention from Mr. Evans, the gardener, for forty years. The largest is in the form of an immense round-topped mushroom, 36 yards round and about 36 feet high, with a perfectly smooth, close, regular surface. In the west garden at the same place there is a double row of yews, eleven on each side, of the same form, but very much smaller.

TIMBER

Since foreign timber has almost entirely superseded home-grown wood, the remarkable qualities of this most durable and beautiful timber have been almost forgotten, though, if we may believe what Evelyn, Loudon, Walker, and other old authors tell us, it was formerly highly valued, not only for bow-making, but for all purposes where strength and durability, when exposed to wet, were required.

At the present time, though I have made many inquiries, I cannot find a cabinetmaker in London who knows or uses the wood; it is rarely to be found in timber yards, and I was told by one of the principal timber merchants in York that I was, in his forty years' experience, the first person who had ever asked for it. It has little or no selling value, and may be bought occasionally for about half the price of oak.

In various old houses, however, examples may be found of its use for furniture, panelling, and inlaying, which show what the wood is worth, when well selected and thoroughly seasoned, to people who do not mind a little trouble.

Evelyn says that for posts to be set in the ground and for everlasting axle-trees there is none to be compared to it, and that cabinetmakers and inlayers most gladly employ it.

Loudon quotes Varennes de Feuilles, who states that the wood, before it has been seasoned and when cut into veneers and immersed some months in pond water, will take a purple-violet colour.

---

[1] Veitch's *Man. Coniferæ*, 137 (1900).   [2] *Gard. Chron.* 1874, p. 264.
[3] *Some English Gardens*, pp. 34, 42, 107 (1904).

Dr. Walker[1] speaks of the yew as a tree which grows well in the shade of rocks and precipices, especially near the sea-shore. " No timber is planted in Scotland that gives so high a price as that of yew and laburnum." He mentions a yew that grew on a sea-cliff, in the small stormy island of Bernera near the Sound of Mull, which, when cut into logs, loaded a large six-oared boat, and afforded timber to form a fine staircase in the house of Lochnell.

Sir Charles Strickland tells me that yew wood which is occasionally dug up in the bogs and fens of East Yorkshire is of a pinkish grey colour, and the most beautiful English wood he knows, but the samples of it which Henry has procured in Ireland are much darker in colour.

Miss Edwards states that in the Pyrenees water vessels are made of yew wood, which have the property of keeping the water cool in hot weather, and that there is a flourishing manufacture of such vessels bound with brass hoops at Osse.

Marshall is quoted by Loudon to the effect that about 1796 yew trees at Boxhill were cut down and sold to cabinetmakers at high prices for inlaying, one tree being valued at £100, and half of it actually sold for £50. Boutcher says that, from his own experience, bedsteads made of yew wood will not be approached by bugs. Mathieu[2] states that in France the wood is sought for by turners, sculptors, and makers of instruments and toys.

The thin straight shoots of the yew which are cut by gipsies in the south of England make most excellent whip sticks, lighter than, and quite as tough as holly. I believe that yew would also make first-rate handles for polo sticks and golf clubs, though makers of these articles do not as yet seem to have used it.

Boulger[3] says that in the library of the India Office there is a Persian illuminated manuscript on thin sheets of yew, and it also makes very ornamental boards for bookbinding.

As an example of what can be done with yew wood, I may refer to Macquoid's *History of English Furniture*, where a coloured illustration (plate iv.) is given of an extremely handsome armchair in Hornby Castle, the property of the Duke of Leeds. Macquoid says :—" The date is about 1550. It is made of yew, which adds to its rarity, for up to this time it was practically penal to employ yew wood for any other purpose than the manufacture of the national weapon ; in this instance the wood has become close, as hard as steel, and of a beautiful dark amber colour."

At Hatfield House, the historic mansion of the Marquess of Salisbury, the small drawing-room is panelled entirely with yew wood, the doors being also made of fine burry pieces, but the workmanship in this case is not perfect, and the colour of the wood has been spoilt by varnish.

At Dallam Tower, Westmoreland, the seat of Sir Maurice Bromley-Wilson, the staircase is made of yew wood grown on the property.

Trees are occasionally found in which the whole body of the log consists of small burry growths something like that of maple, and when this is mixed with contorted grain of various shades of pink the effect is very good. But such trees

---

[1] *Economical History of the Hebrides*, vol. ii. pp. 205, 240 (1812).
[2] *Flore Forestière*, 511 (1897).   [3] *Wood*, 346 (1902).

are even now so valuable that they are cut into veneer, and I have a magnificent specimen of such in a sheet 8 feet long by 18 to 20 inches wide which has been mounted for me as a table by Messrs. Marsh, Cribb, and Co. of Leeds.

The reason why it is neglected for all these purposes is apparently as follows :— The tree is usually grown in the form of a bush, and does not often become tall and straight enough to form clean timber. It is not usually planted close enough to become drawn up into clean poles, and is rarely felled except when in the way, or when it has become decayed and unsightly.

No tree is so deceptive in appearance as an old yew tree. Not only is it usually full of holes and shakes, but the heartwood is generally more or less unsound when over a foot in diameter. Some defects are usually present in an old yew tree, and even when clean and sound, the heartwood is not so good in colour as the younger wood or the slabs; and as the bark grows over and covers all these defects it is generally impossible to say how much, if any, of the timber of a large yew will be useful until it is sawn through the middle.

It seems to be soundest and best in colour when of moderate age and not over 12 to 18 inches in diameter, though the slabs from old trees of which the heartwood is pale, shaky, or faulty often show the finest and most twisted grain.

(H. J. E.)

CRYPTOMERIA

Cryptomeria, D. Don, *Trans. Linn. Soc.* xviii. 166 (1839); Bentham et Hooker, *Gen. Pl.* iii. 428 (1880); Masters, *Jour. Linn. Soc.* (*Bot.*) xxx. 23 (1893).

A GENUS with one very variable living species, in Eastern Asia, belonging to the tribe Taxodineæ of the order Coniferæ.

A tree with evergreen leaves spirally arranged and decurrent on the shoots, which are only of one kind. Flowers monœcious. Male flowers: spike-like, sessile in the axils of the uppermost leaves of the branchlets, composed of numerous imbricated stamens, which have a pointed connective, and 3 to 5 pollen sacs. Female flowers: globular cones solitary and sessile on the tips of branchlets near to those on which the staminate flowers occur, composed of numerous bracts with free recurved pointed ends spirally imbricated in a continuous series with the leaves. Ovular scales, each bearing 3 to 5 ovules, united with the bracts for three-fourths of their length and dilated into roundish crenately-lobed extremities. Fruit: a globular brownish cone, ripening in the first year, but persisting on the tree after the escape of the seeds by the gaping apart of the scales till the next year or longer; scales about 20 to 30 in number, peltate, stalked with a disc dilated externally, which shows on its outer surface the recurved point of the bract (incorporated with the scale in its greater part), and on its upper margin 3 to 5 sharp-pointed rigid processes. The stalk-like portion of the scale bears on its inner side 2 to 5 seeds, which are ovate-oblong, somewhat triquetrous in section, and narrowly winged, with a mucro near the apex.

CRYPTOMERIA JAPONICA

Cryptomeria japonica, Don, *Trans. Linn. Soc.* (*Bot.*), xviii. 167, tab. xiii. 1 (1839); Hooker, *Icon. Plant.* vii. 668 (1844); Siebold, *Flora Japonica*, ii. 43, tab. 124, 124*b* (1870); Kent in *Veitch's Man. Coniferæ*, 263 (1900); Shirasawa, *Iconographie des Essences Forestières du Japan*, text 24, tab. ix. 25-42 (1900); Mayr, *Fremdländische Wald- und Parkbaüme*, 278 (1906).

A tall tree, attaining in Japan a height of 150 feet or more, and a girth of 20 to 25 feet, the trunk tapering from a broad base. Bark reddish brown, and peeling off in long, ribbon-like shreds. Leaves persistent for 4 or 5 years, arranged spirally on the shoots in five ranks, curving inwards and directed forwards, awl-shaped, tapering to a point, compressed laterally, keeled on front and back, bearing stomata on both sides, with the base decurrent on the branchlet to the insertion of the next leaf. The buds are minute, and composed of three minute leaves, which are free at the base, and not decurrent.

The male flowers are clustered at the ends of the branchlets in false racemes, the leaves in the axils of which they arise being reduced in size, and fulfilling the function of bracts. They appear on the tree in autumn and shed their pollen in early spring, remaining for some time afterwards in a withered state.

The buds of the female flowers are also to be seen in autumn terminating some of the branchlets, and covered externally with small, awl-shaped leaves.

The shoot[1] is frequently continued in the leafy state throughout the cone ("proliferation"), and the extended portion often grows to several inches in length beyond the cone, and even in some cases bears male catkins.

Woody excrescences[2] of a conical shape often develop on the stem, to which they are loosely connected. They correspond to the "wood-balls" which are found on beeches and cedars, and like these are due to abnormal development of dormant buds.

Seedling: the cotyledons, which are generally 3 in number, the occurrence of 2 only being rare, are carried above ground by an erect caulicle, about $\frac{1}{2}$ inch long, ending below in a primary root, which is reddish, flexuous, and about 3 inches long, giving off a few lateral fibres. The cotyledons are linear, flattened, obtuse, and about $\frac{1}{4}$ inch long; two, narrowed at the base, are prolonged on the caulicle as ribs; the other, sessile on a broad base, is not decurrent; all bear stomata on their upper surface. The first leaves on the stem are in a whorl of 3, similar in shape to the cotyledons, but longer and with slightly decurrent bases. The leaves following are inserted spirally on the stem, and are longer, sharper-pointed, and more decurrent. All are spreading, with stomata and a prominent median nerve on their lower surface. The stem, roughened by the leaf-bases, terminates above in a cluster of 5 to 6 leaves, crowded at their insertion and directed upwards.

---

[1] Remarkable instances of proliferous cones and other abnormalities are described and figured in *Rev. Horticole*, 1887, 392.

[2] Figured in *Gard. Chron.*, May 30, 1903, p. 352.

Varieties

There are at least two well-marked geographical forms, var. *japonica* and var. *Fortunei*, which will perhaps be ranked as distinct species, when the trees are studied in the wild state. Other varieties, which have probably arisen in cultivation, are distinguished by peculiarities of the foliage.

1. Var. *japonica*, the type described by Don from Japanese specimens collected by Thunberg.—This is the form which occurs wild in Japan. The tree is pyramidal in habit, with straight, spreading branches and short, stout, dark green leaves. The cones are composed of numerous scales, bearing long acuminate processes, and showing long points to the bracts, making the outer surface of the cone very spiny, especially towards the summit. There are generally 5 seeds to each scale.

2. Var. *Lobbii*.[1]—Tree narrow, pyramidal in habit, with short branches densely ramified. The leaves are long and light green in colour. The cones are like those of the preceding variety, but with the processes and tips of the bracts even longer and more slender. This is perhaps a geographical form, occurring in Japan, where it was collected by Wright. It has certainly proved hardier than the Chinese variety both in this country and on the Continent.

3. Var. *Fortunei*[2] or *sinensis*.[3]—A tree diffuse in habit, with deflexed branches and long, slender branchlets. Leaves long and slender. Cones with fewer scales (about 20), which end in short processes, the tips of the bracts being of no great length, so that the whole cone looks much less spiny than that of the Japanese forms. Seeds fewer, often only 2 on a scale, but apparently indistinguishable from those of the Japanese trees. This is the form which occurs wild in China, and which was first introduced into this country. It was described by Sir W. J. Hooker[4] from specimens gathered by Sir Everard Home in Chusan. The Chinese form ripens its seeds three weeks sooner at Dropmore than the var. *Lobbii*.

4. Var. *araucarioides*.[5]—Branches deflexed, with the branchlets long, pendulous, and very distantly placed. Leaves small, stout, stiff, and curving inwards at the top, dark green in colour. Cones as in var. *japonica*, of which this is only a slight variety. It is described as a shrub or low tree; but this may arise from its being propagated from cuttings. Large trees occur, of a similar habit, which seem, however, to be sports from var. *Fortunei*.

5. Var. *pungens*.[6]—Leaves straight, stiff, spreading, darker green, and more sharply pointed than in common forms. I have not seen cones; and the origin of this variety is not clearly known.

---

[1] Gordon, *Pinetum* (1858), p. 54.

[2] *Cryptomeria Fortunei*, Hooibrenk, *Wien. Jour. für Pflenzenkunde*, 1853, p. 22.

[3] *Cryptomeria japonica*, var. *sinensis*, Siebold, *l.c.* 49.

[4] Hooker, *loc. cit.* He points out the difference in the cones of the Chinese and Japanese trees, but says that they are undoubtedly one species. Bunbury, *Arboretum Notes*, 172, remarks that the cones of the Barton tree, from Chinese seed, are very different from Don's figure of Japanese cones.

[5] Carrière, *Traité Gén. Conif.* (1867), p. 193.

[6] *Hort.* A sub-variety of this, *pungens rubiginosa*, is mentioned in *Garden*, iii. 1873, p. 322. The leaves are said to assume a coppery or tawny red colour from August until April.

I   S

6. Var. *spiralis*.[1]—A slender shrub, with leaves strongly falcate and twisted spirally by their free ends around the branchlets, which assume in consequence a corkscrew-like appearance. A specimen of this at Kew also bears some branchlets with normal leaves.

7. Var. *dacrydioides*.[2]—Leaves very closely set and very short (about $\frac{1}{4}$ inch long) There is a specimen at Kew of this form, gathered by Maries and said to be wild. It is probably a depauperate form, originating in rocky, barren, exposed ground.

8. Var. *nana*.[3]—A dwarf, procumbent, dense, spreading shrub, with short acicular needles, closely set on the rigid branchlets and directed outwards. This form attains only 3 or 4 feet in height, and very often bears monstrous fasciated twigs.

9. Some slightly variegated forms of Cryptomeria have appeared in cultivation ; in one the tips of the branchlets are whitish ; in another the leaves are yellowish in colour.

10. Var. *elegans*, Masters, *Jour. Linn. Soc. (Bot.)* xviii. 497 (1881) ; *Cryptomeria elegans*, Veitch, ex Henkel und Hochstetter, *Synopsis der Nadelhölzer*, 269 (1865).— A fixed seedling form. The juvenile foliage is retained throughout the life of the tree, which bears the same relation to the type as *Retinospora squarrosa* does to *Cupressus pisifera*. It agrees in cones and in the anatomical structure of the leaves with the typical form.

In habit this is rather a large bush than a tree. The leaves, while spirally arranged on the shoot as in the ordinary form, spread outwards and are not directed upwards. They are decurrent on the branchlets, linear, flattened, curving downwards, sharp-pointed, grooved on the middle on both surfaces, and are light green in colour, changing in late autumn and winter to a reddish bronze colour, which gives the tree a remarkable and handsome appearance. There is a dwarf form of this variety, *Cryptomeria elegans nana*, which is a low, dense bush with crowded leaves, changing in colour in the autumn like the ordinary variety, except that the pendulous tips of the branchlets remain green.

The origin of this remarkable form is obscure. In Japan, according to Siebold, it is known as *to-sugi*, *i.e.* " Chinese Cryptomeria," and is said to have been introduced from China. Kaempfer mentions a *nankin-sugi*, introduced into Japan from China, cultivated on account of its beauty, which is possibly this variety.

Cryptomeria elegans was introduced from Japan to England in 1861 by John Gould Veitch.[4] The largest specimen we know occurs at Fota ; it is 42 feet high by 4 feet 9 inches in girth. In Cornwall this variety grows to a great size, the tops of the trees often bending down under the weight of their branches and foliage ; and the outer lower branches commonly take root and grow into independent trees, which form a colony round the parent stem.[5]

At Tregothnan there is a very fine example (Plate 37) which measures 35 feet by 4 feet 6 inches, and at Killerton there is another almost equal in size. In the

---

[1] Siebold, *loc. cit.* 32.   [2] Carrière, *Traité Gén. Conif.* (1867), p. 193.
[3] Knight, *Syn. Conif.* (1850), p. 22.
[4] Veitch, *Man. Coniferæ*, 1st ed. 218 (1881) : " Met with only in cultivation in neighbourhood of Yokohama."
[5] *Jour. Hort. Soc.* xiv. (1892), p. 30.

pinetum at Cowdray this form also grows very well, and it is perfectly hardy at Colesborne and in Yorkshire. At Poltalloch in Argyllshire it also attains large dimensions.

IDENTIFICATION

Cryptomeria resembles Sequoia and *Araucaria Cunninghami* in having leaves which are spirally arranged and markedly decurrent on the shoots. The awl-shaped leaves of *Araucaria Cunninghami* strongly resemble those of the ordinary forms of Cryptomeria; but in the former they always end in bristle-like points, whereas in the latter they taper to a blunt point. The subulate leaves of *Sequoia gigantea* are closely appressed to the shoots in three ranks, with only their upper half free; whereas in Cryptomeria they are in five ranks, and are free from the shoots for the greater part of their length.

INTRODUCTION

The tree is said by Siebold[1] to have been introduced into St. Petersburg by the overland route through Siberia, several years before Fortune sent it to England. The credit of the introduction into England is, however, due to Captain Sir Everard Home,[2] who sent seeds to Kew from Chusan in 1842. Several seedlings were raised at Kew, which were kept in a greenhouse till 1847, and were then planted out; but they never did well. One planted near the rockery was living in 1880, when it measured 26 feet high by 2 feet 3 inches in girth; and another stood for some years near the main entrance. Both these trees have been cut down, and there do not appear to be any survivors of the first importation now at Kew or elsewhere. Fortune introduced the tree in quantity in 1844, when he sent seeds, apparently gathered in Chekiang, from Shanghai to the Horticultural Society. The first tree planted in France was at Chaverney in 1844, and the second at Angers in 1847. All the old trees in this country and on the Continent are from Fortune's seeds, and belong consequently to the Chinese form.

The variety *Lobbii* was introduced by Thomas Lobb in 1853 from the Botanic Garden of Buitzenborg in Java, where it had been sent from Japan in 1825 by Siebold.[3] It differs only slightly from the ordinary Japanese form. Siebold[4] states that he introduced the typical Japanese form into Leyden in 1861. John Gould Veitch introduced several kinds of Cryptomeria, as the result of his visit to Japan in 1860; but I have not been able to identify these, and according to H. J. Veitch,[5] the typical Japanese form was first introduced by Maries in 1879. Probably there are no trees of this kind in England older than this date. The introduction of variety *elegans* has been already given above. (A. H.)

DISTRIBUTION IN CHINA

Cryptomeria was discovered in China in 1701 by J. Cunningham, who found it in

[1] Siebold, *loc. cit.* 48.

[2] John Smith, *Records of Kew Gardens* (1880), p. 289; Sir W. J. Hooker, *Guide to Kew Gardens*, 1847, p. 28, 1850, p. 14.

[3] Siebold, *loc. cit.* 48. [4] *Ibid.* 51. [5] *Jour. Hort. Soc.* xiv. (1892), 30.

the island of Chusan, off the coast of Chekiang.   His specimens, three in number, are preserved in the British Museum, and a branch with cones was figured both by Petiver [1] and Plukenet.[2]   A few years previously, probably in 1692, it had been found in Japan by Kaempfer,[3] whose specimen also is kept in the British Museum ; and Thunberg [4] obtained specimens in Japan, the material on which Don founded his description of the genus.   It was collected in China at different times by Sonnerat,[5] Millett,[6] and Sir Everard Home ;[7] but we owe to Fortune the only account of importance of the tree in the wild state in China.

Fortune [8] saw the tree for the first time in the plain of Shanghai in 1843, where it is planted in cemeteries and temple-grounds, and grows to a great size, the poles which are set up in front of temples and mandarins' offices being often of Cryptomeria. In 1844 he found it growing wild in the mountains south-west of Ningpo, where it forms dense woods with Cunninghamia and other trees.   It is met with in the Chekiang mountains even at high elevations—the finest specimens seen by Fortune occurring in the Bohea hills, which he crossed in 1849.   He was particularly struck with a fine solitary tree, at least 120 feet in height, which stood near the gate in the pass of the high range which separates the Chekiang and the Fukien provinces. At Ningpo the junks are mostly built of the timber of this tree.

Cryptomeria was also collected by Swinhoe [9] in the country inland from Amoy, and by Père David in the interior of Fukien, where, he says,[10] it is a beautiful tree, becoming rare in the wild state, but existing still in the mountains at moderate elevations.

Specimens [11] have been collected in other parts of China, but always, I believe, from planted trees.   In Yunnan I only met with two trees, one (18 feet in girth) near a temple, and the other near a village.   Cryptomeria apparently only occurs wild in China in the mountains of the Chekiang and Fokien provinces, between 25° and 29° N. lat., but may be found elsewhere when the interior of the country is better explored.   In its native home in China the tree is subject to severe cold in winter, but the spring arrives suddenly with no late frosts, and the summer is much warmer than in England.   In China the tree is called *kuan-yin-sha* ("goddess of mercy fir") in Yunnan, *kung-ch'io-sung* ("peacock-pine") in Szechuan, and *sha* ("fir") simply in Chekiang, where it shares the name with Cunninghamia, the timber of both trees being much used in the construction of houses and boats.

CRYPTOMERIA IN JAPAN

Both from an ornamental and economic point of view this is the most important

---

[1] Petiver, *Gazophylacium Naturæ et Artis* (1702), tab. 6, fig. 3, "*Cupressus chusanensis, Abietis folio,* from Chusan"; and *Phil. Trans.* xxiii. (1703), p. 1421, No. 70.

[2] Plukenet, *Amaltheum Botanicum* (1705), text 69, tab. 386, fig. 3.

[3] *Amœnitates Exoticæ* (1722) p. 883.   Kaempfer's figure is published in *Icon. Kaempf.* (1791), t. 48.

[4] *Flora Japonica*, 265 (1784).   [5] In 1776.   Lamarck, *Ency. Bot.* ii. 244.

[6] Collected at Macao, where the tree is only planted.   Hook, *Ic. Plant.* vii. t. 668 (1844).

[7] Sir Everard Home collected specimens (British Museum and Kew) in Chusan and near Woosung, and his notes say "from trees near tombs and joss-houses."

[8] Fortune, *Residence among the Chinese* (1857), pp. 145, 184, 189, 256, 277, 412.

[9] Specimen at Kew.   [10] *Plantæ Davidianæ,* i. 291 (1884).

[11] By Anderson, near Momien in Yunnan, and by Dr. Faber, on Mount Omei in Szechuan.

tree of Japan, as it is also the largest, and though it is now difficult to say how far its natural distribution extends, it has been planted everywhere from such a remote period, and grows so rapidly that it is now the most conspicuous tree in all those parts of Japan which I visited except in the island of Hokkaido.[1]

I saw it wild in the primeval forests which cover the mountains on the frontier of the provinces of Akita and Aomori in the extreme north, near a station called Jimba, at an elevation of about 1000 feet, where the lower edge of the forests and more accessible valleys have already been denuded of their best timber. The Japanese Government have lately made a good road up one of the valleys, which enabled me to see the forest at its best under the guidance of their obliging foresters. The hills are here very steep, often with a slope of 30° to 40°, and covered on the north-east aspects with an almost pure growth of Cryptomeria, and though on the south-west aspects a few deciduous trees, such as maple, magnolia, oak, chestnut, and Æsculus were mixed with it, I saw no other conifer. This forest is not truly virgin, because from time to time trees have been cut for shingles and tub staves, which are made in the forest and carried out on men's backs as usual in the remoter parts of Japan. But in many places it was quite dense, and the undergrowth consisted largely of ferns, Aucuba, Skimmia, Hydrangea, and a variety of other shrubs, and tall, rank-growing herbaceous plants such as Spiræa and *Rodgersia podophylla.*

The trees average in size 100 to 110 feet high by 2 to 3 feet in diameter, and are clean for half their length or more, in the denser parts of the forest. The largest trees which have been felled here do not exceed about 100 feet in timber length and about 4 feet diameter. The rings of one of 5 feet in girth which I measured showed 116 years' growth, of which about 87 were red heart-wood. Another close by was very flat-sided, measuring 3 feet 9 inches in diameter one way, and only 2 feet 9 inches the other, the centre on that side being only 1 foot from the nearest point of the bark. This tree was about 136 years old, over 100 years growth being red heart-wood.

Many trees were more or less curved at the butt, and many others forked low down into two, three, or more stems. There were plenty of cones on the trees which had sufficient light, but a careful search did not discover a single self-sown seedling, all the young trees which were coming up—and those not numerous—being evidently suckers or growths from the stool. The dense layer of coarse, sour humus and half-decayed leaves and branches form a bed in which the seedling after germination cannot take root, but on the railway banks and other exposed surfaces not overgrown by dense grass young seedlings appeared and grew freely. Many of these trees had large climbers, such as Vitis Coignetiæ, Schizophragma, and Wistaria, growing nearly to their tops. Plate 38, taken from a negative kindly given me by the Japanese Imperial Forest Department, shows the appearance of this forest. Plate 39 A, from the same source, shows a mature forest of Cryptomeria in the island of Shikoku. Plate 39 B shows the trunk of the tree and the manner of felling still adopted in Japan, *cf.* p. 137.

The forester told me that the system adopted in this forest, now that it is

[1] Mayr quotes Dr. Honda for the fact that the most southern locality where it grows wild is in the island of Yakushima, the northernmost of the Riu-Kiu islands, where in a dense forest at a high elevation it forms immense trunks.

accessible, would be clean felling, followed by replanting as soon as possible, in the same manner as is generally adopted in the south of Japan.

I could not learn the exact range of Cryptomeria as a wild tree,[1] but in the north, where the winter is long and hard, and the snow lies deep for months, it prefers the shady aspect, though it does not attain the same gigantic proportions as it does farther south.

Nikko is approached by a magnificent avenue of Cryptomerias on both sides of the road, 20 miles long, known more or less imperfectly by every visitor to that place, but which can only be properly appreciated by going some way east of Imaichi station, to the point where the trees in good soil attain their greatest dimensions. I took a photographer here specially to take the picture reproduced, and measured the finest trees I could find, of which the tallest was about 145 feet high, and the average 110 to 120 feet, with a girth of 12 to 20 feet on the better soils. Many of the trees have been planted so close together that they have now grown into one tree. The one which I figure (Plate 40) is composed of six stems, which measure 21 feet in greatest diameter, and about 60 in girth. *Cf.* Sargent, *Forest Flora of Japan*, p. 75.

The age of these trees, of which many have been blown down by recent gales and some felled, is, as near as I could count the rings of wood, 260 to 270 years, of which over 200 is red wood. The bark is not over $\frac{1}{2}$ to $\frac{3}{4}$ inch thick, and though some of the trees were beginning to decay at the heart, others were quite sound. The soil is generally a rich black humus overlying a yellow tufaceous volcanic gravel, and the influence of bad soil on the trees is seen very clearly at a point about three miles east of Imaichi, where the road crosses a low ridge of dry and sandy soil, and where they are not more than 80 to 90 feet high by 6 to 8 feet girth.

At the celebrated temples of Nikko there are larger trees than any that I saw in the avenue. The best—shown in Plate 41—is about 150 feet high by 23 feet in girth, but I could not measure the height exactly on account of its position. They are said to be about 300 years old, being probably older than those in the avenue, and seem mostly in perfect health on a slope facing south where the soil is evidently deep and good.

But these magnificent trees are quite eclipsed by those which I saw later at the celebrated monastery town of Koyasan, in the province of Kishu, not nearly so well known to European tourists as it should be. The magnificent cemetery at this place is over a mile long, and planted as an irregular avenue with many lateral annexes— each of which was in the past the private burying ground of great families—with Cryptomeria trees which are said to be 400 years old, and which, I believe, surpass in grandeur any other trees planted by man in the world. They grow at an elevation of about 2800 feet, in a climate which is much milder, and gives evidence of a much heavier rainfall than that at Nikko; for many of the trees had shrubs growing on them as epiphytes on their trunks. In one case a tree of *Cupressus obtusa* has its stem, 6 to 8 inches thick, completely embedded in the trunk of a sound and

---

[1] In *Forestry of Japan*, p. 18, it is only said that splendid natural pure woods of it occur in the Nagakizawa State forests in Akita, and in Yakujima in the island of Kyushu, which I had not time to visit, but whether there is any notable difference between the trees in these distinct areas, separated by nearly ten degrees of latitude, is not stated, so far as I can find. According to Shirasawa (*loc. cit.*) fossil Cryptomeria trees of great dimensions have been found in nearly all parts of Japan

healthy Cryptomeria, from whose sap alone it must now be deriving its sole nourishment, as no decaying wood is visible, and it is about 20 feet from the ground. The shape of the trees here is more picturesque and less regular than at Nikko, some having spreading branches quite near the ground; the best of these measured 133 feet by 19 feet 3 inches, with a spread of 25 yards.

The finest trees in the cemetery, and probably the finest in existence, stand on the right at its extreme end, close to an enclosure, just before reaching the large barn-like temple called " Mandoro," or hall of ten thousand lamps, which is itself surrounded and backed up by a grove of superb trees standing very thickly together. Of the trees on the right just before reaching the temple one had previously been measured by Prof. Honda of Tokyo University, who made it 58 metres high. I made it 180 feet with a girth of 24 feet. But though this may be the tallest it is not so fine a timber tree as the one standing just beyond it, which does not swell so much at the ground, but carries its girth higher up and is cleaner. This tree is broken off at about 150 feet, but seems quite vigorous, and certainly contains 2000 feet or more of sound timber.[1]

So far I have spoken only of the Cryptomeria in a wild state and as an ornamental tree, but it is also planted very largely in many parts of Japan for timber, and forms a most profitable source of revenue to many of the smaller landowners and farmers as well as to the State. Its cultivation has attained a maximum in the district of Yoshino in the province of Yamato, and from *The Forestry and Forest Products of Japan*, published at Tokyo in 1904, we learn that this cultivation dates back 400 years, and covers as much as 38 per cent of the whole area of the district, of which no less than 93 per cent is forest land. The inhabitants have probably brought the art of profitable timber growing to a higher point of perfection than any other people in the world, no less than 85 per cent of the local male population consisting of woodmen, sawyers, timber carriers, and foresters. The quantity of Cryptomeria timber alone exported from Yoshino amounted in the year 1902 to 8,857,000 cubic feet, valued (I presume locally) at 1,695,000 yen, equal to about £175,000 sterling.

The trees are planted out at three years old after being twice transplanted in the nursery, where they are raised from seed and kept shaded during the first year. This, at least, is the rule in the Kisogawa district, though I was told that in the south Cryptomerias are more cheaply and quickly raised from cuttings, and that these produce as good trees as seedlings.

About 4000 per acre are usually planted, and weeded once or twice a year for three years, when they suppress the weeds by their shade. The plantations grow very fast, and are pruned from the eighth to the twenty-third year after planting out. Thinning is done at the earliest at twelve years, and the thinnings form such a profitable source of revenue that income is probably returned quicker by such a Cryptomeria plantation than by any other tree. The final felling takes place at about 120 years old, when as many as 180 trees, containing 15,000 cubic feet, may be found on an acre. The previous thinnings are estimated at 16,000 cubic feet, making the total product per acre in 120 years over 30,000 feet. This result, which

[1] Mayr, however, states that he measured a tree at Takaosan which attained 68 metres (over 200 feet) in height by 2 in diameter.

appears astonishing, is perhaps exceptional, but all the plantations I saw gave evidence of extremely rapid growth, and showed a larger proportion of clean useful poles and timber than any plantations which I have seen in other countries.[1]

TIMBER

The wood is used for almost every purpose in Japan, but especially for tubs, staves, and building. Though not as valuable as the best wood of *Cupressus obtusa* for high-class buildings and internal work, it is, when properly selected, sawn, and planed, highly ornamental both in colour and grain, easy to work, durable, and strong enough for most purposes. It has also a most agreeable odour due to the presence of a volatile oil called *sugiol* by Kimoto,[2] who gives an analysis of it, and states that the wood on this account is used for making *saké* casks, the *saké* acquiring a peculiar pleasant aroma.

It varies very much in colour and figure, the most valuable being the wide planks—sometimes 3 to 4 feet wide and over—which are used for doors, ceilings, and partitions in the best houses. The darkest in colour comes from the southern island of Kiusiu, and is known as *Satsuma sugi*. When it shows a very fine red grain in old gnarled butts it is known as *Ozura-moko*, the best of this colour being very valuable. There is also a grey-coloured variety known as *Gindai sugi*, which appears to be taken from trees which have died before felling, but I could not get very definite information on this point.

The finest example I know of the ornamental use of Cryptomeria wood is the ceiling of the large dining-room in Kanaya's Hotel at Nikko, which is composed of panels about 30 inches square, cut from the butts of trees which show very curly and intricate graining, and without polish have a natural lustrous gloss. The Japanese never paint or varnish the wood in their houses inside or out, and attach more importance than European builders do to its quality, colour, and figuring. It seems very strange that none of the numerous travellers and writers on Japan have, so far as I can learn, as yet paid any attention to the beauty of the Japanese timbers. As a rule Cryptomeria is spoken of by English-speaking Japanese and Europeans as cedar, but *sugi* is the native name.

The bark of the tree is also largely used, when taken off in large sheets, for covering outbuildings of secondary importance, but does not appear to be so much valued or so durable as the bark of *Cupressus obtusa*, *Thujopsis dolabrata*, or *Sciadopitys verticillata*. An ounce of the seed contains about 50,000 seeds. For raising trees to plant in the colder parts of Europe I should certainly prefer seed from the natural forests of the north to what is grown in the subtropical climate of South and Central Japan, and I should therefore warn anyone wishing to plant this tree largely to be very careful about the origin of the seed or plants.

The value of this wood varies considerably in Japan according to locality and

[1] Tables of Production and Rate of Growth in Japan are given by Honda in *Bull. Coll. Agric. Tokyo Imp. Univ.* ii. 335 (1894-1899).   [2] *Bull. Coll. Agric. Tokyo Imp. Univ.* iv. 403 (1900-1902).

quality, but about 80 yen per 100 cubic feet, equal to about 1s. 8d. per foot, is the price in Tokyo, and selected half-inch boards for ceilings and panellings cost from 2s. to 4s. each.

Rein, in *Industries of Japan*, p. 226, speaks of a Cryptomeria which he measured in 1874 having at 1½ metres high a girth of 9.41 metres, equal to about 30 feet. This grew on the Sasa-no-yama-toge, between Tokyo and Kofu, at about 750 metres above sea level.

Weston, in *The Alps of Japan*, mentions trees high up on the eastern side of the pass between Nakatsugawa and the Ina-kaido, called the Misaka-tōge, on the northern slope of Ena-San, which measured at 3 feet from the ground no less than 26 feet in girth. It would not be supposed possible that in a country where neither machinery nor horse-power is used for the removal of timber such large trees could be utilised, but the Japanese are very ingenious in the handling of large logs in their mountain forests.

I was presented by Baron Kiyoura, Minister of Agriculture, with a most curious and interesting series of sketches, which I found in the Imperial Bureau of Forestry, showing the means adopted for felling and transporting large timber growing in rocky gorges and the most inaccessible situations. These I exhibited at a meeting of the Scottish Arboricultural Society at Edinburgh on 10th February 1905.

The *modus operandi* is as follows :—First men climb up the trees and lop off all the large branches, so that the tree may not lodge among its standing neighbours when felled. Ropes are then attached to the trunk and carried round a windlass, so that it may be pulled over in the right direction.

When the tree is felled it is cut into suitable lengths, often 20 to 30 feet long, and a hole cut in the end, to which a stout rope is attached. By this it is sometimes dragged, sometimes lowered, to the nearest slide, which is built up of smaller timber. Or, if the locality is too distant from a slide or from a stream large enough to float it, a platform is built on the mountain-side, on which it is sawn into boards, which are carried out of the forest on men's backs, or on sledges on the snow.

A most ingenious plan, which I have seen in no other country, is adopted where the slope of the mountain is too steep to let a log slide at its own pace.

The slide is built in a zig-zag form, and at each angle a bank is made and covered with earth and bark to check the impetus of the log, whose upper end when so checked is reversed by means of a strong pole laid across the slide, and then goes downwards till it reaches the next angle, where it is again checked and reversed by its own weight.

To see a large gang of men, all singing in chorus at their work, moving timber in a mountain forest under the direction of their foremen, is one of the most interesting sights I beheld in Japan, and I could not sufficiently admire the pluck, activity, and ingenuity they showed in the very dangerous and difficult work which is necessary when logs get jammed, as they often do in these rapid mountain torrents ; and when men, standing on small rafts fastened to boulders in a roaring rapid, or let down from above by ropes, have to dislodge the logs from where they have stuck fast.

Cryptomeria japonica yields in Japan a turpentine or semi-solid resin, named *sugi-no-janai*, which was shown at the Edinburgh Forestry Exhibition. This resin, which is very aromatic, is used as incense in Buddhist temples, and as a plaster for wounds and ulcers.

CULTIVATION

The tree ripens seeds in good summers in the south of England, which are easy to raise, and the seedlings grow rapidly after the first year. Seeds which I gathered in 1900 from a tree at Stratton Park, Hants, the seat of the Earl of Northbrook, and sowed in the open ground, germinated at the end of April, whilst others sown in a pot on Christmas Day and kept in a greenhouse, germinated on 17th April, and grew much better than those sown in the open. Some of the young trees planted out at two years old are now (January 1905) 3 to 4 feet high, and have not suffered at all from the spring frosts.[1]

The seedlings are easy to transplant, and might be raised in nurseries at a lower rate than many trees, though they should have some protection for the first two or three years, and if kept in pots the roots should not be allowed to become cramped, and if twisted round the bottom of the pot should be carefully spread when planted out. The tree may also be propagated from cuttings, and this plan is sometimes adopted in Japan, as being cheaper and quicker than raising seedlings, but except in the case of varieties, should not be adopted if tall, straight trees are desired. I have seen in the garden of Mr. Chambers at Grayswood, near Hazlemere, Sussex, a self-sown Cryptomeria which had germinated in a chink of the garden steps, and which is now growing at Colesborne, and I have no doubt that others might be found in suitable situations, as Mr. Bartlett has lately found a seedling at Pencarrow, Cornwall, growing at the base of the parent tree. Cryptomeria seems to be more adaptable to various kinds of soil than many exotic trees, and does not mind a moderate amount of lime, but loves a situation well sheltered from cold winds, and a soil deep enough and light enough to keep its roots moist during summer. I have not seen it grow well on heavy clay, where it suffers from spring frost. If timber and not ornament is the object, I should plant it about 10 feet apart, alternate with some fast-growing conifer, such as common larch or spruce, which might be cut out when the trees became too crowded.

Mayr[2] considers that in warm, damp parts of Europe the Cryptomeria may probably be planted profitably as a timber tree in sheltered valleys and in good soil, but recommends the mixture of other trees as nurses wherever the winter temperature is low, and says that alders are preferred by the Japanese for this purpose. He says that a plantation of this tree in East Friesland had attained 12 metres in height and 23 centimetres in diameter; and on the island of Mainau, on Lake Constance, he measured, in 1897, a tree which was 18 metres high and 40 centimetres in diameter.

---

[1] But the severe frosts of May and October 1905 have injured several and killed some of the weakest of these seedlings.

[2] *Fremdländische Wald- und Parkbäume*, p. 285.

REMARKABLE TREES

The finest specimens of Cryptomeria known to us in England are at Hempsted, the seat of the Earl of Cranbrook, in Kent, which Lord Medway thinks were planted before 1850. They grow in a sheltered situation on a greensand formation, and the largest exceeds *Sequoia sempervirens*, planted near it, in height. I made it 80 feet by 8 feet, and another tree 72 feet by 8 feet 2 inches. Both are symmetrical, and seem to be growing fast (Plate 42).

At Pencarrow, in Cornwall, a number of trees were planted by Sir W. Molesworth in 1848-9, and have thriven very well, though the soil is not very favourable. Mr. Bartlett informs us that the tallest of them, which grows on a dry, steep, stony bank, crowded by other trees, is now 68 feet by 5 feet 6 inches. The largest, on moister soil, is 62 feet by 8 feet. Six other trees, planted in 1875, vary from 48 feet by 5 feet down to 33 feet by 4 feet 3 inches. At this elevation, 450 feet, there is often snow, and the thermometer sometimes falls to 12° and 14°.

At Castlehill, in North Devon, the late Earl Fortescue planted many Cryptomerias, but though they have grown well, they are mostly rather bushy than tall trees, and may have been raised from cuttings. One of them, which was 8 feet in girth and only 40 feet high, was covered with burr-like excrescences as much as 8 inches long.

At St. George's Hill, near Byfleet, growing on Bagshot sand on the highest part of the hill, surrounded by pines, is a tree 64 feet by 5 feet 5 inches.

At Kitlands,[1] near Leith Hill, Surrey, there is a large Cryptomeria, planted by the late D. D. Heath, Esq., the branches of which have taken root and formed a grove, whose branches in turn root outside.

There is a fine tree at Killerton, which in 1902 was about 75 feet high, by 6 feet 5 inches in girth, though owing to its situation it was difficult to photograph or to measure exactly. A tree at Bicton is nearly equal to it in height and girth. At Eastnor Castle, Worcestershire, the property of Lady Henry Somerset, there is a tree 65 feet high by 5 feet 10 inches in girth.

At Fonthill Abbey, Wilts, the seat of Lady Octavia Shaw-Stewart, there is a beautiful tree 67 feet high by 9 feet 3 inches in girth. At the entrance gate of Rufford Abbey, Notts, the seat of Lord Savile, there is a fine tree about 62 feet high.

At Dropmore a large tree, planted in 1847, was cut down in 1904,[2] and measured 68 feet 6 inches by 5 feet 9 inches. Though it seemed in perfect health, Mr. Page informed me that the heart was partly decayed. There are still three good specimens at Dropmore—two of the Chinese kind, planted in 1847, which measure (February 1905) 64 feet by 5 feet 6 inches, and 62 feet by 6 feet 7 inches; the third, var. *Lobbii*, planted in 1853, is 74 feet high by 4 feet 7 inches in girth. These measurements were kindly sent by Mr. Page.

At Barton, Suffolk, one, planted in 1848, was found by Henry in 1904 to be

---

[1] Nisbet in *Victorian Surrey County History*, ii. 575 (1905).
[2] *Gard. Chron.* Jan. 21, 1905, p. 44. This tree was reported to be 41 feet high in 1868 (*l.c.* 1868, p. 464).

63 feet by 5 feet 3 inches, and proves that the tree will thrive on good soil even in the east of England.

At Williamstrip Park, Gloucestershire, the seat of Lord St. Aldwyn, there is a tree 45 feet by 5 feet, probably one of the first introduced.

At Penrhyn, North Wales, there is a fine tree, 64 feet by 5 feet 5 inches.

The best example I know of the growth of Cryptomeria as a forest tree is at Tan-y-bwlch, in Merionethshire, the property of W. E. Oakeley, Esq., where a large number of seedlings were raised about 40 years ago from a tree which is now 62 by 6 feet, and has apparently not grown much lately.

The best of its progeny, growing on a slate formation where rhododendrons flourish exceedingly, near sea level, is already 53 by 4½ feet, and many others average about 40 by 3½ feet.   Some are growing among beech and oak, others in a plantation of larch and Corsican pine facing north.   In the latter the average girth of 8 trees was 3½ feet, whilst larch of the same age was little over 2 feet, and Corsican pine about the same.   Mr. Richards, forester to Lord Penrhyn, who saw this plantation shortly after it was made, agreed with me that its success would amply justify planting Cryptomeria on a large scale in North and West Wales in sheltered places on good land up to about 300 feet above the sea.   But, judging from a large board sent me by Sir John Llewellyn, grown in South Wales, the timber is much lighter and softer than it is in Japan, and perhaps will not be equal for outside work to that of Douglas fir grown on similar land.

At Dynevor Park, in Caermarthenshire, the seat of Lord Dynevor, there are some well-grown trees, the tallest of which is 56 feet by 6 feet 10 inches.

At Belshill, Northumberland, the property of Sir W. Church, there is a tree 50 feet by 4 feet 8 inches, which is about 50 years old and quite healthy.

In Scotland the tree seems quite hardy, and at Keir, the seat of A. Stirling, Esq., a tree planted in 1851 has increased from 42½ feet in 1892 to 52 feet by 8 feet girth in 1905.   It has the trunk covered with burrs.   At Castle Kennedy, the seat of the Earl of Stair, there is a tree 56 feet by 6 feet 1 inch.

In Ireland there are several fine trees.   At Coollattin, Wicklow, the property of Earl Fitzwilliam, a tree measured 63 feet high by 6 feet in girth in 1906.   At Woodstock, Kilkenny, a tree of the variety *Lobbii*, which was planted by Miss Tighe in 1857, is now (1904) 67 feet high by 6 feet 7 inches in girth.   Close by this tree is a wonderful group of Cryptomerias, which have been produced by natural layering. The parent tree in the centre is about 50 feet high, and around it are over 20 trees, with straight stems, which are themselves layering, so that in course of time a grove may be produced.

At Fota there are many examples of Cryptomeria and its varieties.   The form *spiralis* is about 15 feet high.   The variety *araucarioides* is 31 feet by 3 feet 8 inches in girth, very compact in habit.   *Elegans* is 42 feet by 4 feet 9 inches.   The variety *Fortunei* measures 72 feet high, with a girth of 8 feet, a beautiful pyramidal tree, displaying the stem below with its characteristic stringy bark.   This tree was planted in 1847.   (H. J. E.)

PYRUS

Pyrus, Linnæus, *Gen. Pl.* 145 (1737); Bentham et Hooker, *Gen. Pl.* i. 626 (1865).
Malus, Ruppius, *Fl. Jen.* ed. 3, 141 (1745); and Medicus, *Phil. Bot.* i. 138 (1789).
Sorbus, Linnæus, *Gen. Pl.* 144 (1737).

TREES and shrubs belonging to the sub-order Pomaceæ of the order Rosaceæ. Branchlets of two kinds, long and short shoots, the flowers in certain species being borne on the latter only. Leaves deciduous, alternate, stalked, simple or pinnate; stipules deciduous. Flowers in cymes or corymbs, regular, perigynous or epigynous, calyx-lobes 5, petals 5. The receptacle (the end of the axis) is hollowed out, the ovary being attached to its interior. A disc is present, either annular or coating the receptacle. Ovary with 2 to 5 cells, each cell containing 2 ovules. Fruit, a pome, the external fleshy part being formed of the receptacle, while the interior or core is the developed ovary; cells 2 to 5, with a membranous or cartilaginous endocarp, each containing 1 or 2 seeds, though occasionally some are empty.

The genus Pyrus has been divided variously into sections, which some botanists treat as distinct genera. The following arrangement is perhaps the simplest :—

A. *Leaf in the bud rolled inwards towards the midrib.*

1. **Pyrophorum.** Flowers in corymbs on spur-like branchlets, ovary with 5 cells, styles free. Fruit pyriform or hollowed out at the base, flesh granular. Leaves simple. Pears: confined to Asia and Europe.

2. **Malus.** Flowers fascicled or umbellate on spurs, ovary with 3 to 5 cells, styles united at the base. Fruit with a cavity at the base, flesh homogeneous. Leaves simple. Apples: species in North America as well as in Europe and Asia.

3. **Aronia.** Flowers in terminal corymbs, ovary with 4 to 5 cells, styles free or united at the base. Fruit small, not hollowed at the base, endocarp very thin, flesh almost homogeneous. Leaves simple, crenate, with the midrib glandular on its upper side. Two North American shrubs.

B. *Leaf folded in the bud. Flowers in terminal corymbs.*

4. **Hahnia.** Ovary with 2 to 3 cells, styles united below. Fruit crowned by the calyx, and having a hard, almost bony endocarp, flesh granular. Leaves simple, lobed. *Pyrus torminalis*, the only species.

141

5. **Sorbus.** Ovary with 3 or 5 cells, styles free. Fruit crowned by the calyx, endocarp membranous or coriaceous. Leaves pinnate. Includes two sub-sections :—

> **Aucuparia**, with 3-celled ovary and small globular fruit, and
> **Cormus**, with 5-celled ovary and large pear- or apple-shaped fruit.

6. **Aria.** Ovary with 2 to 5 cells, styles free. Fruit crowned by the calyx, endo-carp membranous, flesh granular. Leaves simple. Includes the whitebeam and its allies.

7. **Micromeles.** Ovary with 2 to 3 cells. Fruit small, globose, umbilicate, endo-carp membranous or coriaceous, calyx-lobes deciduous. Leaves simple. Includes several Asiatic species.

SYNOPSIS OF THE PRINCIPAL SPECIES IN CULTIVATION EXCLUSIVE OF PEARS AND APPLES.

(Cf. Plates 43-45, where leaves of most of the species are shown.)

I. *Leaves regularly pinnate, the leaflets being separate and never decurrent by their bases on the rhachis.*

   A. **Aucuparia.** Mountain ashes. Leaflets unequal-sided at the base. Fruit small, not exceeding $\frac{1}{3}$ inch.

   (1-3) *Winter buds white-tomentose.*

   1. **Pyrus Aucuparia,** Gaertner. Europe, Northern Asia, Japan. Young branchlets and leaves pubescent, adult leaves glabrous or only slightly pubescent beneath. The common mountain ash.

   2. **Pyrus lanuginosa,** DC. South-eastern Europe. Only differs from the preceding in the adult leaves being densely woolly beneath.

   3. **Pyrus thianschanica,** Regel. Chinese Turkestan. Young branchlets and leaves glabrous; adult leaves quite glabrous beneath and con-spicuously veined on the upper surface.

   (4-5) *Winter buds shining, glutinous, glabrous or sparingly pubescent, the pubescence appressed and of a rusty colour.*

   4. **Pyrus americana,** Torrey and Gray. North America. Leaflets long, narrow, acuminate, glabrous beneath.

   5. **Pyrus sambucifolia,** Chamisso and Schlechtendal. Manchuria, North-East Asia, Japan, North America. Leaflets broader than in No. 4, obtuse or acute (not acuminate), more or less pubescent beneath.

   B. **Cormus.** True Service. Leaflets nearly equal sided at the base. Fruit large, $\frac{1}{2}$ inch diameter or more.

6. **Pyrus Sorbus**, Gaertner. Central and Southern Europe. Winter buds greenish, viscid, pubescent only at the tip; under surface of the leaves slightly pubescent in spring, soon becoming glabrous.

II. *Leaves pinnate, but upper 3 or 5 leaflets coalesced or decurrent by their bases on the rhachis.*[1]

7. **Pyrus Aucuparia**, Gaertner, var. **satureifolia**, Koch. A hybrid. Differs from the common form in the coalescence into one large segment of the 3 upper leaflets; leaflets glabrous beneath.

8. **Pyrus Aucuparia**, Gaertner, var. **decurrens**, Koehne (*Pyrus lanuginosa*, Hort. *non* DC.) A hybrid. Upper leaflets decurrent on the rhachis by broad bases, often the upper 3 or 5 coalescing into one segment; leaflets tomentose beneath.

III. *Leaflets pinnate or deeply cut at the base, with 1-4 pairs of segments, the upper part of the leaf lobed or serrate; leaves very variable in shape.*[1]

9. **Pyrus hybrida**, Moench. A shrub of hybrid origin. Main axis of the leaf glandular above, under surface of the leaf sparingly pubescent, the parents being *Pyrus aucuparia* and *Pyrus arbutifolia*.

10. **Pyrus pinnatifida**, Ehrhart. A hybrid. Axis of the leaf without glands, under surface densely grey tomentose.

IV. *Leaves simple, lobed.*

A. *Under surface of the leaf glabrous, or nearly so, light green in colour.*

11. **Pyrus torminalis**, Ehrhart. Europe, Algeria, Asia Minor, and the Caucasus.

B. *Under surface of the leaf grey tomentose.*

12. **Pyrus cratægifolia**, Savi. Italy. Leaves small, resembling those of a hawthorn, on each side 4-6 triangular-ovate toothed lobes.

13. **Pyrus latifolia**, Boswell Syme. Britain, France, Spain, Central Europe. Leaves broad-oval with a wide base; lobes decreasing from below upwards, small triangular, separated by sinuses, which form a right angle and are not narrowed. In some forms the lobes are mucronate, in others cuspidate; and, in var. *decipiens*, the outline of the leaf is like *intermedia*, but the lobing is different.

14. **Pyrus intermedia**, Ehrhart. Europe. Leaves elliptic, with a usually narrow base, lobes decreasing from below upwards, rounded, mucronate, separated by narrow sinuses, which are very acute or almost closed at their bases. This includes several forms :—

Mougeoti. Leaves with 9-12 pairs of nerves, lobes shallow.

[1] These two sections comprise hybrids, the leaves of which vary in shape, not only on different individual trees, but often also on a branch. Hybrid origin may always be suspected when such variation is observed, or when the lobing or cutting is irregular and not symmetrical.

Scandica. Leaves with 7-9 pairs of nerves, lobes deep, with sharp
teeth.

Minima. Leaves with 5-7 pairs of nerves, smaller and narrower
than in the preceding varieties.

Certain forms of *Pyrus pinnatifida* closely resemble *scandica*, but the
lobing in these will be found always irregular and often very deep.

15. **Pyrus lanata,** D. Don. Himalayas. Leaves large, broad oblong,
woolly underneath, nerves 12-15 pairs, lobes regularly serrate.

C. *Under surface of the leaves, which are orbicular in outline, snowy-white
tomentose.*

16. **Pyrus Aria,** Ehrhart, var. **flabellifolia.** Greece. Leaves with 3-5 pairs
of nerves.

17. **Pyrus Aria,** Ehrhart, var. **græca.** Greece, Asia Minor. Leaves with
6-10 pairs of nerves.

V. *Leaves simple, not lobed, and only occasionally obscurely lobulate.*

A. **Aronia.** *Leaves finely serrate in margin, with glands on the upper surface of
the midrib.* This section comprises 2 North American species and a hybrid
of garden origin, small shrubs, only referred to here to prevent their being
confused with other species of Pyrus.

18. **Pyrus arbutifolia,** Linnæus fil. North America. Leaves beneath
whitish grey tomentose, with about 6 pairs of nerves directed forwards
at a very acute angle.

19. **Pyrus alpina,** Willdenow. A hybrid between *Pyrus Aria* and the
preceding species. Leaves densely grey tomentose beneath, with
9-10 pairs of nerves directed outwards at an angle of 45°.

20. **Pyrus nigra,** Sargent. North America. Leaves glabrous beneath or
very slightly pubescent.

B. *Leaves without glands on the midrib.*

(21-22) *Leaves glabrous beneath.* These 2 species, of which the first is a shrub
and the other a small tree, are only referred to here to distinguish them from
other species.

21. **Pyrus Chamæmespilus,** Linnæus. Vosges, Jura, Alps, Pyrenees.
Leaves sessile or nearly so, elliptic, with 6-8 pairs of nerves.

22. **Pyrus alnifolia,** Franchet and Savatier. Japan and China. Leaves
stalked, broadly ovate, with 9-12 pairs of nerves.

(23-26) *Leaves white pubescent beneath.*

23. **Pyrus Aria,** Ehrhart. Europe, Caucasus, Siberia, Central China.
Leaves oval or elliptic with very slight lobules or only doubly-
toothed, the teeth or lobules diminishing in size from above downwards; nerves 7-12 pairs, very prominent on the lower surface,
pubescence snowy white.

24. **Pyrus Aria**, Ehrhart, var. **rupicola.**   British Isles (Europe ?) Leaves obovate - oblong (broadest above the middle), lobulate above, the lobules and teeth acute, nerves 5-9 pairs, pubescence at first as white as in the type, but ultimately becoming greyer.

25. **Pyrus Aria**, Ehrhart, var. **Decaisneana.**   Origin unknown.   Leaves large, elliptic, or oblong, with margin serrated almost uniformly, nerves 12-15 pairs.

26. **Pyrus Hostii**, Hort.   A hybrid.   Leaves like 23, but with very sharp, irregular teeth and tomentum *thin*, white to greyish white.

(27) *Leaves grey, densely-woolly pubescent beneath.*

27. **Pyrus vestita**, Wallich.   Himalayas.   Leaves very large, elliptic, serrate, and occasionally obscurely lobulate in margin ; nerves 15-18 pairs.

As many of the species mentioned above are merely shrubs or very small trees, they do not fall within the scope of our work.   For this reason, *Pyrus hybrida, cratægifolia, Chamæmespilus, alnifolia*, and the three species of the section Aronia, will not be further referred to.   *Pyrus Hostii*, a hybrid of inconstant origin, will be briefly mentioned in connection with *Pyrus Aria.*

Pyrus Aucuparia and its allies will be dealt with in a subsequent part.

The two following species are not known to us to attain timber size in cultivation in the British Isles ; but Mr. H. C. Baker tells us that at Chilternhouse, near Thame, there is a specimen of *P. vestita* 50 feet by 6 feet 5 inches.

PYRUS LANATA, Don.[1]

Known in gardens as *Sorbus majestica.*   A tree of the eastern temperate Himalayas ; leaves large, oval, oboval, or broadly oblong, with serrate lobes, glabrous above when adult except for some pubescence along the midrib, greyish woolly beneath.   Flowers white in densely woolly corymbs ; petals glabrous within ; styles 2 to 5, free, densely tomentose.   Fruit large, about an inch in diameter ($\frac{1}{2}$ to $1\frac{1}{2}$ in.), narrowed to the base, red, edible.   Judging from wild specimens the foliage is very variable ; and the cultivated specimen at Kew bears leaves (figured in Plate 43), which differ from those of wild trees in being less deeply lobed.

PYRUS VESTITA, Wall.[2]

Often known in gardens as *Sorbus nepalensis* or *Sorbus magnifica.*   A tree of the temperate Himalayas from Garwhal to Sikkim.   Leaves (*cf.* Plate 43) very large, ovate-acute or elliptic, lobulate-serrate, densely covered with white wool when they first appear, but later in the season becoming shining green and glabrous above, remaining densely woolly beneath.   Flowers in very woolly corymbs ; petals woolly within ; styles 3-5, tomentose only at the base.   Fruit large, about $\frac{3}{4}$ inch in diameter, globose.

---

[1] Don, *Prodromus*, 237 (1825).   Hook., *Fl. Brit. Ind.* ii. 375 (1879).
[2] Wallich, *Catalogue*, 679 (1828).   Hook., *Fl. Brit. Ind.* ii. 375 (1879).

PYRUS SORBUS, TRUE SERVICE[1]

Pyrus Sorbus, Gaertner,[2] *De Fruct*. ii. 43, t. 87 (1791), Loudon, *Arb. et Frut. Brit*. ii. 921 (1838).
Pyrus domestica, Ehrhart,[2] " Plantag," 20, ex *Beiträge zur Naturkunde*, vi. 95 (1791) ; Smith, *Eng.
    Bot*. t. 550 (1796).
Sorbus domestica, Linnæus, *Sp. Pl*. 477 (1753).
Cormus domestica, Spach, *Hist. Vég. Phan*. ii. 97 (1834).

A tree, attaining a height of 60 to 80 feet.   Bark, like that of the common pear,
dark brown, fissuring longitudinally, and scaling off in narrow, rectangular plates.
Leaves pinnate : 6 to 9 pairs of sessile leaflets and a terminal stalked leaflet.   Leaflets
linear oblong, almost equal-sided at the base, and acute at the apex, serrate with
acuminate teeth, except towards the base where they are entire ; dull green above,
paler below, glabrous on both surfaces when mature, some pubescence often, however,
remaining underneath.   Flowers white, in short pubescent corymbs ; styles 5,
united at the base and woolly in their whole length.   Fruit either pear- or apple-
shaped, generally green, tinted with red on one side, 5-celled, about an inch in
diameter.

The fruit apparently varies much in flavour, but in good varieties is agreeable
though astringent.   The French proverb, *Ils ne mangent que les cormes*, applied to
destitute persons, would indicate that the fruit was poor ; and this is doubtless often
the case.   In parts of France a perry is made from them, and they are also
preserved dry like prunes.   At Vevay[3] in Switzerland there are avenues planted,
consisting of service trees of various kinds ; and the brilliancy of the fruit and of
the hues of the foliage in October give a very fine effect.

VARIETIES

Two well-marked forms occur, one *maliformis*,[4] with apple-shaped fruit, the other
pyriformis,[4] with pear-shaped fruit.   There would seem, however, to be in France,
though little known to planters in general, varieties which produce fruit of a superior
kind.   Two of these are strongly recommended by a writer in the Journal of the
French National Horticultural Society :[5] one discovered on the estate of M. Dufresne,
near Bordeaux, which has large pyriform fruits of a carmine yellow, produced in large
bunches and excellent in flavour, as soon as they commence to mellow ; the other
was also found growing wild in woods belonging to M. Lafitte at Agen, which has
fruit of a bright pink colour.

[1] *Service* is commonly derived from the Latin *cerevisia*, a drink said to have been formerly made of berries of the
different species of Sorbus, or to have been flavoured with their leaves.   C. Woolley Dod controverts this view in *Gard. Chron.*
1890, vii. 87, and holds that *service* is simply a corruption of *sorbus*, and that *cerevisia*, a drink, according to Pliny, made of
cereal grain in Gaul, was ordinary malt ale.

[2] Gaertner's and Ehrhart's names were both published in the same year.   Gaertner's preface antedates that of Ehrhart
by a few days.   Nothing is known for certain of the pamphlet " Plantag " cited by Ehrhart.   Which name has priority of
publication is uncertain.

[3] *Woods and Forests*, July 16, 1884.   [4] Loddiges, *Catalogue*, ex Loudon, *loc. cit.*
[5] Quoted in *Garden*, 1886, xxx. 89.

IDENTIFICATION

In summer the tree is only liable to be confused with the mountain ash and its allies. The bark is, however, different, being rough, scaly, and dark-coloured in the true service tree, smooth and grey in *Pyrus Aucuparia*, etc. In *Pyrus Sorbus* the leaflets at the base are practically symmetrical, and their serration is very acute. Buds if present are the best distinction, as explained below.[1] In winter *Pyrus Sorbus* is distinguished by the following characters, shown in Plate 45 :—

Twigs: long shoots glabrous, round; leaf-scars, crescentic with 5 bundle dots, set parallel to the twig on projecting cushions. Terminal buds larger, side buds coming off at an acute angle; all ovoid, densely viscid, shining, generally pubescent at the tip. Bud scales few in number, greenish, sometimes reddened, viscid, quite glabrous, the margin without cilia. Short shoots ringed, glabrous, ending in a terminal bud. The viscid greenish buds, 5-dotted leaf-scars, and rough scaly bark, distinguish this species from other kinds of Pyrus.

DISTRIBUTION

The Service tree is largely cultivated in central and southern Europe; and in many places, where it is recorded as wild, is really only an escape from cultivation. It is met with in the forests of France which rest upon limestone; but in the north and east it does not produce fruit every year, and is doubtfully wild except in the south and west. Willkomm considers it to be wild in the southern parts of the Austrian empire (Dalmatia, Croatia, Banat, Carniola, and South Tyrol), in the valley of the Moselle, in the Jura and Switzerland; also in southern Europe and Algeria. In France it is occasionally met with as a standard in coppiced woods.

Mouillefert says that the tree may live to be 500 or 600 years old, and that it was uninjured by the severe frost of 1879-80, when the thermometer fell to $-25°$ Reaumur. He says, also, that it prefers a rich calcareous soil, but will grow on sand if not too dry. (A. H.)

REMARKABLE TREES

Pyrus Sorbus is not a native of Britain, though a single specimen which grew in a remote part of Wyre Forest in Worcestershire was long considered to give it a claim to be introduced into the British flora. This tree was mentioned in the *Philosophical Transactions*[2] as long ago as 1678 by Mr. Pitt, who says that he found it in the preceding year as a rarity growing wild in a forest of Worcester, and identifies it with the *Sorbus pyriformis* of L'Obelius, a tree not noticed by any preceding writer as a native of England. Pitt says nothing about the size of the tree, merely observing : " It resembles the Ornus or quicken tree, only the Ornus bears the flower

---

[1] The stipules of the various species of the section Sorbus differ considerably in shape, as shown in Plate 43 ; but they are usually quickly deciduous, and can only assist identification in spring.

[2] *Phil. Trans.*, abridged edition, ii. 434 (1809).

and fruit at the end, this on the sides of the branch.   Next the sun, the fruit has a dark red flush, and is about the size of a small jeneting pear.   In September, so rough as to be ready to strangle one.   But being then gathered and kept till October, they eat as well as any medlar."

Ray's account[1] in 1724 is as follows : " The true Service or Sorb.   It hath been observed to grow wild in many places in the mountainous part of Cornwall by that ingenious young gentleman, Walter Moyle, Esq., in company with Mr. Stevens of that county.   I suspect this to be the tree called *Sorbus pyriformis*, found by Mr. Pitt, alderman of Worcester, in a forest of that county, and said to grow wild in many places of the morelands in Staffordshire by Dr. Plot, *Hist. Nat. Stafford*, 208."   In modern times the tree has, however, never been found wild in any part of Cornwall or Stafford, and probably it was confused with *Pyrus latifolia*.[2]

Nash,[3] in 1781, refers to the tree in the Wyre forest as occurring " in the eastern part of Aka or Rock parish, about a mile from Mopson's Cross, between that and Dowles Brook, in the middle of a thick wood belonging to Mr. Baldwyn, which I suppose to be the *Sorbus sativa pyriformis*, mentioned by Mr. Pitt in the *Philosophical Transactions for* 1678, called by the common people the Quicken pear tree."   This tree was figured by Loudon,[4] t. 644, from a drawing sent him by the Earl of Mountmorris.   The Rev. Josiah Lee, rector of the Far Forest, told Mrs. Woodward that the old inhabitants of the district, where it was called the " Whitty Pear tree," used to hang pieces of the bark round their necks as a charm to cure a sore throat.   Lee's *Botany of Worcestershire*, 4, gives a good figure of this tree " from a sketch taken many years ago," and another as it appeared in 1856; and says that in 1853 it was in a very decrepit state, producing a little fruit at its very summit.   It was burnt down in 1862, by a fire kindled at its base by a vagrant.   In a note Lees says that he thinks the tree must have been brought from Aquitaine and planted beside a hermitage in the forest, of which no trace is left but a mound of stones overgrown by brambles.   He found the privet and *Prunus domestica* occurring near it, and nowhere else in the forest.

A seedling (Plate 46) from the Wyre forest tree is growing on the lawn at Arley Castle, near Bewdley, formerly the property of Lord Mountmorris, now the residence of Mr. R. Woodward.   I measured it in 1903, when it was 55 feet high by 7 feet 4 inches in girth, and quite healthy, though a large hole in the trunk has been filled with cement.   A few seedlings have been raised from it at Arley, but grow very slowly.

There is a large healthy tree in the park at Ribston Hall, Wetherby, the seat of Major Dent, of the pyriform variety, which in 1906 I found to be about 65 feet high by 9 feet in girth, and bearing fruit.   This tree was probably brought from France by the same Sir. H. Goodricke who sowed the original Ribston pippin in 1709.

---

[1] Ray, *Synopsis Methodica*, ed. 3, p. 542.

[2] Miller, *Gard. Dict.* iii. ed. (1737), under *Sorbus sativa*, says, " The manured service was formerly said to be growing wild in England ; but this I believe to be a mistake, for several curious persons have strictly searched those places where it was mentioned to grow, and could not find it ; nor could they learn from the inhabitants of those countries that any such tree had ever grown there."   [3] *History of Worcestershire*, i. 11.

[4] Loudon gives the measurements in 1838 as 45 feet high, with a diameter of trunk at a foot from the ground of 1 foot 9 inches, and states that it was in a state of decay at that time.

At Croome Court, the seat of the Earl of Coventry, there are two good-sized trees in the shrubbery, one of which is 59 feet high and 6 feet 2 inches in girth. The other, with a clean stem, about 50 feet by 7, is beginning to decay.[1] Lady Coventry told me that the fruit, which is only produced in good seasons, makes excellent jam when fully ripe, but some seeds which she was good enough to send me did not germinate.

Loudon mentions a tree at Melbury Court, Dorsetshire, estimated to be 200 years old, and 82 feet high, with a diameter of 3 feet 4 inches, growing in dry loam on sand. If this was really a true sorb, it must have been the largest on record, but I learn from the gardener at Melbury that it has long been dead.

There are two good-sized trees at Painshill, and another at Syon which Henry found to be 44 feet high and 6 feet 9 inches in girth, but on this heavy soil the tree does not seem to be so long lived, and is dying at the top.

In the Botanic Gardens at Oxford are two well-shaped trees of this species, which were laden with fruit in 1905, and supposed to have been planted by Dr. John Sibthorp, who was Professor of Botany in 1784-95. The largest measures about 50 feet by 5 feet, and is of the maliform variety. Its fruit, which ripens and falls about the middle of October, is very sweet and pleasant to eat, much better than medlars, whilst the fruit of the other, which is the pyriform variety, does not turn red, is smaller, and ripens later. I have raised seedlings from both of these trees.

In the Cambridge Botanic Garden there is a tree with very upright branches, which measured, in 1906, 42 feet by 3 feet 4 inches.

At Tortworth there is a healthy, well-shaped tree, not more than 40 years planted, which is about 40 feet by 5 feet 11 inches. This is in a rather exposed situation, and it had no fruit in 1905.

At Woodstock, Co. Kilkenny, Ireland, there is a tree which seems to be the largest now living in this country. Henry measured it in 1904 and found it 77 feet high by 10 feet 8 inches in girth, with a bole dividing into three stems at 10 feet from the ground and bearing fruit.

TIMBER

A large tree was blown down at Claremont Park, Surrey, the seat of H.R.H. the Duchess of Connaught, in 1902, which I am assured by Mr. Burrell, the gardener there, was a sorb.[2] Its trunk was sent to Mr. Snell of Esher, to whom I am indebted for two fine planks of its wood. These show a very hard, heavy, compact surface of a pinkish brown colour with a fine wavy grain, which takes an excellent polish, and this wood has been used with beautiful effect in the framing

[1] Loudon speaks of a tree at Croome 45 years planted, and 80 feet high, which is possibly the same, but his measurements are very unreliable.

[2] "Among interesting trees to be found at Claremont is a good specimen of the pear-shaped service, carrying a heavy crop of fruit. It is rather over 60 feet high and 7 feet 6 inches in girth at 2 feet from the ground." Note by E. B. in *Garden*, 1883, xxiv. 422. Mr. E. Burrell gives a fuller account of the Claremont trees in the same journal, 1888, xxxiii. 154, in which he states that he thinks the variety *maliformis* does not increase in height after it gets to be about 30 feet high, whereas *pyriformis* at Claremont is close on 70 feet high.

of a brown oak chest, made for me by Messrs. Marsh, Cribb, and Co., of Leeds. Mouillefert says it is one of the hardest and most valuable woods grown in France, and is especially sought for by engravers, carvers, turners, and gun-makers.   It seems to be difficult both to propagate and to grow, at least in its youth, and Loudon says that though it may be grafted on the pear or the mountain ash, it is one of the most difficult trees to graft, and that it will not layer successfully, and that it grows very slowly from seed, not attaining more than 1 foot high in four years.   Seeds[1] sown in autumn germinate in the following spring.   The young seedling has two oval entire-margined cotyledons, and attains about 4 inches in height in the first year's growth.   Plants may be had from the French nurserymen.   Mr. Weale, of Liverpool, reports as follows on a sample of this wood which I sent him :—" The wood is close and homogeneous in texture, tough, but inclined to be brittle.   Rays on transverse section invisible, and rings only to be distinguished by the difference in colour of the spring and autumn wood.   Harder than whitebeam, seasons well, without warping or splitting, and with little shrinkage.   A reliable wood when thoroughly dry."   (H. J. E.)

[1] Mathieu, *Flore Forestière*, 184 (1897).

PYRUS TORMINALIS, Wild Service

Pyrus torminalis, Ehrhart, " Plantag." xxii. ex *Beit. zur Naturkunde*, vi. 92 (1791) ; Loudon, *Arb. et Frut. Brit.* ii. 913 (1838) ; Conwentz, *Beob. über Seltene Waldbäume in West Preussen*, 3 (1895).
Cratægus torminalis, Linnæus, *Sp. Pl.* 476 (1753).
Sorbus torminalis, Crantz, *Stirp. Austr.*, ed. 2, fasc. ii. 85 (1767).
Torminaria Clusii, Roemer, *Synopsis*, iii. 130 (1847).

A tree, attaining exceptionally a height of 80 feet, but more generally only reaching 40 or 50 feet. Bark smooth and grey at first, but after fifteen or twenty years of age scaling off in thin plates, and ultimately becoming fissured. Leaves long-stalked, broadly oval, nearly as broad as long, with a cordate or truncate base and an acute apex ; with 6-10 triangular acuminate serrate lobes ; shining and glabrous above, obscurely pubescent beneath ; nerves pinnate, 5-8 pairs. Flowers white in corymbs. Styles 2, glabrous, united for the greater part of their length. Fruit ovoid, brownish when ripe, with warty lenticels, vinous in taste when in a state of incipient decay ; cut across transversely it shows a ring of white hardened tissue, forming a mesocarp around the core.

The leaves are generally described as glabrous on the under surface, but in all specimens traces of pubescence may be observed, which is much more marked on coppice shoots and epicormic branches.

VARIETIES

None have been obtained in cultivation so far as we know, and wild trees vary very little in any of their characters. A variety, *pinnatifida*, with the lobing of the leaves very deep, is described by Boissier,[1] from specimens occurring in Asia Minor and Roumelia.

IDENTIFICATION

The leaves in summer are unmistakable (see Plate 44), and can only be confounded with certain forms of *Pyrus latifolia* ; but in the latter species the under surface of the leaf is always plainly grey tomentose, and the lobes are much shorter than in *P. torminalis*. In winter the following characters, shown in Plate 45, are available.

Twigs: long shoots, glabrous, shining, somewhat angled, with numerous lenticels ; leaf-scar semicircular with 3 bundle traces, set parallel to the twig on a greenish cushion. Buds almost globular, terminal larger, side-buds nearly appressed to the twig ; scales green with a narrow brown margin, glabrous, with the apex double-notched. Short shoots slightly ringed, glabrous, ending in a terminal bud.

(A. H.)

[1] *Flora Orientalis*, ii. 659 (1872.)

Distribution

The most complete and recent account of the tree is a monograph cited above with maps by Prof. H. Conwentz, who describes at great length the various places where the tree is found and the conditions under which it grows. It is widely distributed throughout most of the woods and forests of Europe, but does not occur in Scandinavia, Holland, or the greater part of Russia, where it is only met with in the southern provinces. It also occurs in the Caucasus, Asia Minor, Syria, and in Algeria. It is found on most geological formations, including granite in the Vosges, gneiss in Siberia, and basalt in Austria; and it prefers a soil rich in humus. Willkomm says that on mountains it is commoner on limestone than on other soils, but the French foresters say that it is practically met with on all soils that are not very dry or very wet. It is a tree of the lowlands and hills, attaining 700 metres altitude near Zurich, 1200 in Herzegovina, and 1900 in the Caucasus. It occurs more or less rarely over all parts of Germany, especially in the north-east, and it attains its maximum size in the royal forest of Osche in West Prussia. The largest tree known to Conwentz was "25 metres high, with a clean stem of 12 metres, and a girth at 1 metre from the ground of 2 metres." The age of this tree was estimated from the rings in the broken trunk of another tree at 235 years.

The scarcity of the tree, as a rule, cannot be accounted for by any deficiency in reproductive powers, for the fruit is produced in some abundance in good years; and being eaten by many birds and animals, among which the waxwing chatterer, the nutcracker, and the fieldfare are mentioned, the seeds must be widely dispersed, while the freedom with which the roots produced suckers is remarked upon.

The timber seems to be much more highly valued in Germany than here, from 18 to 52 marks per cubic metre being given for it, according to quality, in places where hornbeam is only worth 11 marks. It is very hard and durable, and takes a fine polish like that of maple.

In the Hartz mountains and Thuringia it is known as "Atlasholz," and is much used and valued for furniture making.

The fruit is not so much valued as formerly, when it was sold in Prague and Vienna in the winter at the market, and also in Wurtemburg, under the names of Häspele, Arlesbeere, or Adlsbeere.

Conwentz says that the Latin name *torminalis* was derived from the Latin word *tormina*, and given on account of the properties of the fruit, to which one of its names in England, "griping service tree," also has reference.

In Upper Alsace a spirit is distilled from the fruit, which tastes something like Kirschwasser.

Distribution and Remarkable Trees in England

Pyrus torminalis does not occur as a wild tree, and is rarely planted in Ireland, Scotland, or the North of England. Its range is from Anglesea and Nottingham southwards. It is known as the service or griping service tree; and in Kent and

Sussex the fruit is called chequers. It is found in woods, copses, and hedgerows, usually on loam or clay, but does not seem to grow on sandy soils. It attains its greatest size in the Midland counties, where it reaches a height of from 50 to 70 feet. It never seems to be gregarious, and though it reproduces itself by seed or suckers, yet being usually looked on as underwood and not allowed to grow up to its full size, does not attract notice, and is unknown except to the most observant woodmen, even in districts where it occurs. In the vale of Gloucester, on the Earl of Ducie's property, there are probably thirty or forty trees of it scattered over a considerable area. The tallest of these, though not the thickest, is in Daniel's Wood, and is figured (Plate 47). This tree was 62 feet by 5 feet 1 inch in 1904, and is still growing vigorously among other trees and underwood. Not far off is another which may grow to as fine a tree.

In the Cotswold Hills the tree is very rare. I only know a single decaying specimen of moderate size in Chedworth Wood, close to the road leading from Withington to Chedworth Downs.

On Ashampstead Common, in Berks, I found one about 65 feet high and 8 feet in girth, crowded among other trees, which had produced a few suckers. At Rickmansworth Park, Herts, are two fine trees growing together by a pond, which are probably planted (Plate 48), and which Henry measured in 1904, when they were 65 feet by 8 feet 3 inches, and 63 feet by 9 feet 1 inch respectively. The largest specimen known to me is at Walcot, Shropshire, growing on a bank in good soil with a wych elm crowding it on one side, but probably planted. It measures as nearly as I could estimate about 80 feet high, and is no less than 8 feet 9 inches in girth.

At Cobham Hall, Kent, a tree planted beside a pond measured 55 feet high by 4½ feet in 1906.

In the woods in Worcestershire, as I am informed by Mr. Woodward, it is not uncommon, but is not looked upon as of any value. There is a tree at Arley 5 feet 1 inch in girth. In Wychwood Forest, Oxon, now nearly all destroyed, I am informed by Mr. R. Claridge Druce, of Oxford, the tree was formerly common enough for its fruit to be collected and sold in Witney market under the name of service berries.

In Cornbury Park, and in the remains of the forest outside it, there are at least six good-sized trees surviving. Of these the largest, just outside the park wall on the south side, is 65 to 70 feet high by 6 feet 6 inches in girth, a well-shaped, vigorous tree, which on 16th October 1905 was covered with unripe fruit. Another, also outside the park, is about 50 feet by 6 feet 10 inches, with a fine clean bole 12 to 14 feet long. I saw no suckers or seedlings near these trees.

In the woods and coppices north and north-east of London, and in Herts, the tree is not unfrequent on clay soil, and Pryor[1] gives several localities for it.

In Epping Forest, Mr. E. N. Buxton tells me that he does not know of more than thirty trees on an area of 3000 acres, growing on heavy gravelly clay. The largest in his grounds is 40 to 50 feet high.

[1] *Flora of Hertfordshire*, 154 (1887).

In Sussex, Sir E. Loder knows it as an uncommon hedgerow tree of no great size, and Mr. Stephenson Clarke, of Borde Hill, also tells me that it occurs there, and more commonly in the Isle of Wight, where, in Bridlesford Copse, Woolton, a wood of about 200 acres, are perhaps two dozen old trees, which differ in appearance from the Sussex ones in assuming a somewhat pendulous habit of branching when well grown. I find no mention of the occurrence of the tree in the New Forest, and the Hon. G. W. Lascelles does not know it there.

We have no record of *Pyrus torminalis* as a planted tree in Scotland, except that the Rev. Dr. Landsborough[1] notes a tree in vigorous health in Bellfield Avenue, Kilmarnock, which was 2 feet 9 inches in girth in 1893. He calls it the English service tree or table rowan, and adds that, in spite of its Latin name, the fruit is pleasant. In Ireland the tree is very rare. Henry saw, however, a fine specimen in 1903, at Adare, Limerick, which measured 53 feet by 5 feet 10 inches.

The fruit is ripe late in October, when it falls, if not previously eaten by birds, and the seeds, which only seem to mature in warm summers, should be sown at once, or kept in sand exposed to the weather and sown in spring, when they will germinate the next year. Seedlings raised by me from seed gathered at Les Barres, France, which were sown 7th July 1902, germinated 9th March 1903, and were on 14th October 1904 1 to 2 feet high. The leaves turn a reddish yellow in autumn, when the tree is decidedly ornamental, though, on account of its slow growth, it does not seem to have any value as a forest tree, and is rarely procurable from nurserymen in this country.

TIMBER

Pyrus torminalis is unknown as a timber tree in the trade owing to its scarcity, and is mentioned by Boulger[2] only as "a small tree, sometimes 30 feet high, with wood practically identical in character and uses with that of the rowan." Stone does not mention it at all, and Marshall Ward, in his edition of Laslett, says nothing worth quoting.

I am indebted to Mr. Stephenson Clarke for a log of the timber, which resembles that of the whitebeam tree, being hard, heavy, and, according to Loudon, weighing, when dry, 48 lbs. per cubic foot.

Mr. Weale, of Liverpool, reports as follows on a sample of this wood which I sent him:—"Of a hardness between true service tree and whitebeam. Rays on transverse section just visible, a little narrower than sycamore, but wood generally exhibits similar characters. Takes a good finish, but this is not lasting, the ring boundaries rising after exposure. Seasons fairly well, shrinks a little, and rather inclined to twist."

Evelyn says that "the timber of the sorb is useful to the joiner, of which I have seen a room curiously wainscotted; also to the engraver of woodcuts, and for most that the wild pear tree serves."   (H. J. E.)

[1] *Annals of Kilmarnock Glenfield Ramblers' Society,* 1894, p. 11.
[2] *Woods of Commerce,* 312.

PYRUS LATIFOLIA, Service Tree of Fontainebleau

Pyrus latifolia, Boswell Syme, *Bot. Exchange Club Report,* 1872-1874, p. 19 (1875).
Pyrus rotundifolia, Bechstein, N. E. Brown in *Eng. Bot.* iii. ed. Suppl. 164 (1892).
Cratægus latifolia, Lamarck, *Flore Française,* ed. i. 486 (1778).
Sorbus latifolia, Persoon, *Syn. Pl.* ii. 38 (1807).

A tree, attaining a height of 60 feet in France, with smooth, grey bark, which becomes fissured at the base in old trees. Leaves broadly oval, with a broad, rounded, or truncate base and an acute apex; margin with small triangular lobes, decreasing in size from the base of the leaf upwards, dentate and mucronate, the sinuses opening between the lobes almost at a right angle. The leaves are firm in texture, shining and glabrous above, tomentose and greyish green beneath, with 6 to 10 pairs of lateral nerves prominent underneath. Flowers in moderate-sized corymbs, never long peduncled. Fruit globular, $\frac{1}{2}$ inch diameter, smooth, reddish, marked with brown dots, flesh edible; containing two cells, one seed in each cell, or more often one cell with one seed, the other cell containing two aborted ovules.

The description just given is drawn up from Fontainebleau specimens; and trees absolutely identical are said to occur in various forests in Seine-et-Oise, Seine-et-Marne, Marne, Aube, and Yonne.

A series of forms,[1] however, occur in the forests of the east of France, in Alsace-Lorraine, Spain, Switzerland, Austria-Hungary, and Bosnia, which differ slightly in the general outline of the leaf and in the colour and marking of the fruit; and these are supposed to be hybrids between *Pyrus Aria* and *Pyrus torminalis,* between which species they oscillate in the characters of the foliage and fruit; whereas, according to French botanists, the tree of Fontainebleau is a true species, as it reproduces itself naturally by seed; and, moreover, one of the supposed parents, *Pyrus Aria,* is not, according to Fliche, wild in the forest of Fontainebleau.[2] However, the differences are trifling; and it is convenient, in the present state of our knowledge, to treat these supposed hybrids as varieties of *Pyrus latifolia.*

Varieties

Var. *rotundifolia* (Bechstein).[3] Leaves broadly oval or suborbicular, sometimes even broader than long, truncate or rounded at the base, sub-obtuse at the apex; lobes obtusely cuspidate.

Var. *decipiens* (Bechstein).[4] Leaves elongated with acute bases, much resembling

---

[1] These may be called, if their hybridity is considered to be established, *Pyrus Ario-torminalis,* Garcke, *Flora von Deutschland,* ed. 17, 207 (1895). Fliche, in Mathieu, *Flore Forestière,* 177 (1897), sums up the question thus :—Fontainebleau tree not a hybrid, near to *Pyrus Aria,* a true species, seed germinating readily and producing natural seedlings ; Lorraine tree nearer to *Pyrus torminalis* than to *Pyrus Aria,* a true hybrid, seeds rarely perfect. Rouy et Fourcaud, *Flore de France,* vii. 22 (1901), suggest that the Fontainebleau tree is a hybrid fixed and behaving as a true species. See also Irmisch in *Bot. Zeitung,* 1859, p. 277.

[2] *Cf.,* however, p. 156, note 2.   [3] *Pyrus rotundifolia,* Bechstein, *Forstbotanik,* 152 and 316, t. 5, 1843.
[4] *Pyrus decipiens,* Bechstein, *loc. cit.* 152 and 321, t. 7.

those of *Pyrus intermedia*, except that the lobes are triangular pointed, and not rounded as in that species, the sinuses never being acute at their bases.

Var. *semilobata* (Bechstein).[1] Leaves oval or elliptic oval, acute at the apex, narrowed at the base, lobes sharply cuspidate.

IDENTIFICATION

In summer the leaves are distinguishable from those of *Pyrus intermedia* by the characters of the lobes and sinuses ; while broad-leaved forms differ from *Pyrus torminalis* in being tomentose beneath, the lobes never being so long as in that species. The tomentum wears off the under surface of the leaf towards the end of the season, and is never so dense or so persistent as in *intermedia*. On Plate 44 figures are given of leaves from wild trees occurring at Symond's Yat (Fig. 9) and Minehead (Fig. 11), and from a cultivated tree at Kew (Fig. 12). In winter a tree cultivated at Kew showed the following characters, represented in Plate 45.

Twigs : long shoots, shining, round, glabrous, except for a little pubescence near the tip ; lenticels numerous as oval prominent warts ; leaf-scars set somewhat obliquely on prominent, often greenish cushions ; crescentic with three bundle dots, of which the central one is the largest. Terminal bud oval, much larger than the side buds, which come off the twigs at a very acute angle, with their apices bent inwards. All the buds are viscid, pubescent at the tip, and composed of oval scales, which are keeled on the back, ciliate in margin, and short-pointed at the tip. Short shoots ringed, slightly pubescent, ending in a terminal bud. In the specimens examined the leaf-scar at the base of the terminal bud had acute lateral lobes not observed in other species of Pyrus ; but these are probably not always present.

DISTRIBUTION

The tree was first discovered in the forest of Fontainebleau,[2] and was described by Valliant[3] as "Cratægus folio subrotundo, serrato, et laciniato." Duhamel du Monceau gave a figure of the leaf in his classic work.[4] The distribution on the Continent of the type, and of the forms allied to it, has been given above.

In England a small tree, of somewhat rare occurrence, grows wild in woods

---

[1] *Pyrus semilobata*, Bechstein, *loc. cit.* 152 and 317, t. 6.

[2] I visited Fontainebleau in 1905 on purpose to see this tree at home, and found only small trees of it in full flower on 14th May. I was informed by M. Reuss, Inspector of Forests at Fontainebleau, that the tree grows scattered only in the part which is called Montenflammé and Mont Merle, where the sand is covered by the calcareous strata of Beaune, so that the tree is evidently peculiar to calcareous formations. Formerly the trees were cut with the underwood, but are now reserved on account of their rarity, as well as the whitebeam and *P. torminalis*, which M. Reuss considers to be indigenous at Fontainebleau, and therefore admits the possibility of their hybridising. The largest tree known to him is on Mont Merle, at the corner of the roads d'Anvers et de l'Echo in the 16th série, and is 40 centimetres in diameter, or about 4 feet in girth at 5 feet from the ground. It is known to the peasants at Fontainebleau as *baguenaudier* or *elorsier*, but is generally termed by French botanists *alisier de Fontainebleau*.—(H. J. E.)

[3] *Botanicon Parisiense*, ed. 3, p. 63 (1727).   [4] *Traité des Arbres*, i. 194, t. 80, fig. 2 (1755).

in Cornwall, South Devon, and Gloucestershire,[1] which is very near to, if not absolutely identical with, the Fontainebleau tree, as some of the specimens have leaves which resemble rather those of the varieties *rotundifolia* and *semilobata*. The South Devon tree produces fertile seed,[2] which has been planted, and the offspring differs in no respect from the wild trees. In English trees the flowers are reported to have a disagreeable odour,[3] and the fruit ripens in the end of October or November. When fully grown, but still hard, it is olivaceous brown in colour, with numerous scattered small brown or grey dots; but when quite mature it becomes reddish. At Minehead in Somerset, the Nightingale Valley and Leigh Woods near Bristol, and at Castle Dinas Bran, Denbigh, the variety *decipiens* occurs.[4] Mr. E. S. Marshall observed a remarkably fine specimen with good fruit on the Conan river in East Ross-shire; but as no other specimen was seen this tree is probably not wild in that locality. The tree in Earl Bathurst's woods near Cirencester has given rise to some difference of opinion. It was identified at Kew as *Pyrus intermedia*; but in the specimens which I have seen the leaves have the triangular lobing and tomentum of *Pyrus latifolia*, and I have no doubt that it is this species.[5] Its foliage is very variable, some leaves being broad, with rounded bases like the type, whilst others have narrowed bases, and approximate in outline to the *decipiens* variety.

REMARKABLE TREES

Pyrus latifolia is seldom planted except in botanical gardens, as at Kew, Edinburgh, and Glasnevin. There are several fine trees at Edinburgh, one of which was figured in the *Gardeners' Chronicle*[6] for 1882, when it was 45 feet high by 5 feet 3 inches in girth. Professor Balfour had the tree measured again in January 1904, when it was 45 feet high by 6 feet 6 inches in girth. A year or two before it was considerably pruned on the top branches, and this probably accounts for it not being higher in 1904 than it was in 1882. Professor Balfour kindly sent me specimens of the Edinburgh trees, which, though they differ slightly, are all referable to *Pyrus latifolia*. He informed me that while the birds eat the fruit off one tree as soon as it is ripe, in another the fruit remains on the tree untouched. The variability of the fruit in this species is remarkable, and points undoubtedly to hybrid origin.

A tree exists at Oakleigh House,[7] near Keynsham in Somerset, which was planted many years ago. (A. H.)

---

[1] "Occurs at Bicknor, Coldwell, and Symond's Yat, which form a single range of wooded limestone rock in West Gloucestershire, about ¾ mile in length."—Rev. A. Ley, *Bot. Exchg. Club Report*, 1893, p. 415. "French Hales" is the name given to this species in Devon, according to Britten and Holland, *Dict. Eng. Plant Names*, p. 194 (1886). They state that the fruits are sold in Barnstaple market. These authors call the tree *Pyrus scandica*, as, at the time they were writing, its identity with the Fontainebleau tree was not established.

[2] Briggs, *Jour. Bot.* 1887, p. 209, and 1888, p. 236.

[3] Briggs, *Flora of Plymouth*, 144 (1880); and Boswell, *Bot. Exchg. Club Report*, 1872-74, p. 20.

[4] *Cf.* N. E. Brown, *loc. cit.* 165. Mr. J. White reports a tree 30 feet high in Leigh Wood (*Bot. Exchg. Club Report*, 1902, p. 45).

[5] Mr. Hickel, Inspecteur des eaux et forêts, who knows the Fontainebleau tree well, and to whom I sent specimens, is of my opinion.

[6] Vol. xviii. 749.　　　　　　　　　　　[7] *Jour. Bot.* 1899, p. 488.

My attention was called by Mr. R. Anderson of Cirencester to a very remarkable tree growing in a part of Earl Bathurst's woods about two miles from Cirencester, known as the Dear Bit. The tree, though it has lost some of its principal branches, is still, as our illustration shows (Plate 49), a very handsome one, and in size exceeds any other of the kind of which we have a record, either in this country or on the Continent. It is, as nearly as I can measure it, about 75 feet high by 11 feet in girth. It grows on dry shallow soil of the Oolite formation, and is close to a ride, which leads me to suppose that it was planted perhaps at the time when the park was laid out. It is near the north-east edge of the wood, and open to the south-west. I have never seen the flowers of this tree, which bears fruit only in favourable seasons near the ends of its uppermost branches, and as the birds are fond of it, and even in good years many of the seeds are immature, I have not until 1904 been able to procure any. A few of these have now produced small plants.

I have been unable to find any self-sown seedlings near this tree, and though there are one or two good-sized *P. torminalis* in another part of the park, probably planted, none of them approach it in size. As to the possible age of this tree, I can only say that the drive on the edge of which it grows has, as I am told by Mr. Anderson, certainly been in existence over 100 years, and the bank was covered with old beech, which were cut in 1892. The tree has become one-sided from the pressure of a beech which until then closed it in on the south-west side, where it is now open. As these beeches were 150 years old or more, the tree may be now from 150 to 200 years old, and it seems very probable that the person who designed this park had seen the tree at Fontainebleau, and introduced it when Oakley Park was planted by the ancestor of the present Earl Bathurst in Queen Anne's reign.

(H. J. E.)

PYRUS INTERMEDIA, Swedish Whitebeam

Pyrus intermedia, Ehrhart, *Beiträge zur Naturkunde*, iv. 20 (1789); Loudon, *Arb. et Frut. Brit.* ii. 915 (1838).

Pyrus scandica, Ascherson, *Fl. des Prov. Brandenburg*, i. 207 (1864).

Pyrus suecica, Garcke, *Fl. Deutschland*, ed. ix. 140 (1869); Conwentz, *Beob. über Seltene Wald-bäume in West Preussen*, 81 (1895).

Sorbus scandica, Fries, *Flora Hollandica*, 83 (1818).

Sorbus intermedia, Persoon, *Syn. Pl.* ii. 38 (1807).

Sorbus Mougeoti, Soyer-Willemet et Godron, *Bull. Soc. Bot. de France*, v. 447 (1858).

Cratægus Aria scandica, Linnæus, *Amœn. Acad.* 190 (1751).

Cratægus Aria suecica, Linnæus, *Sp. Pl.* 476 (1753).

A shrub or small tree attaining a height of 20 to 50 feet. Leaves stalked, oval or elliptic, rounded or cuneate at the base, pointed at the apex; margin lobed, lobes diminishing in size from the base upwards, rounded, toothed, shortly acuminate, separated by sinuses which are very acute or almost closed at their bases; upper surface green, shining, glabrous when adult, lower surface greyish tomentose. Flowers in branching corymbs, with pleasant odour; petals spreading, tomentose; styles 2, free, tomentose at the base. Fruit oval, red, sweet-flavoured, smooth or slightly dotted.

Varieties

1. *Scandica.*[1] Leaves less narrowed and almost rounded at the base, deeply lobed, with numerous sharp teeth; 6-8 pairs of nerves. Fruits large, surmounted by the curved and outwardly-reflected calyx teeth.

2. *Mougeoti.*[2] Leaves narrowed at the base, slightly lobed, with few short teeth; nerves 9-12 pairs. Fruit very small, surmounted by erect and inwardly-curved calyx teeth.

3. *Minima.*[3] Leaves linear-oblong, with 3-4 pairs of lobes, variable in size, but generally deepest at the middle part of the leaf; nerves 6-8 pairs. Flowers—early in June—in loose corymbs, not flat-topped, small, and resembling those of *Pyrus Aucuparia.* Fruit small, globose, bright red, surmounted by erect calyx lobes.

In Plate 44 figures are given of the leaves of var. *scandica* from Bergen (Fig. 19), of the variety from Great Doward in Hereford (Fig. 10), and of var. *minima* from Breconshire (Fig. 17).

Identification

In summer the greyish tomentum of the leaves underneath, and the rounded lobes, with sharp sinuses which are almost closed at their bases, will distinguish the

---

[1] *Sorbus scandica*, Fries, *loc. cit.*   [2] *Sorbus Mougeoti*, Soy.-Will. et God. *loc. cit.*

[3] *Pyrus minima*, Ley, *Jour. of Bot.* 1895, p. 84, and 1897, p. 289, t. 372; *Sorbus minima*, Hedlund, *Kon. Sv. Veten. Akad. Handl.* (1901-2) 60,

tree from *Pyrus Aria* and *Pyrus latifolia*, the species nearly allied. In winter the following characters are available, as shown in Plate 45 :—

Twigs : long shoots, round, glabrous, often with waxy patches ; lenticels long, numerous. Leaf-scars : crescentic, with 3 equal-sized bundle dots, obliquely set on a brownish projecting cushion. Buds glistening, pubescent at the tip ; terminal much the largest ; side-buds arising at an acute angle with their apices directed in-wards. The bud-scales have a dark-coloured rim to the ciliate margin, and their apex is scalloped with a central projection ending in a tuft of long hairs. The short shoots are ringed, pubescent, with a terminal bud.

DISTRIBUTION

The variety *minima* occurs only in Breconshire, on the limestone mountain cliff Craig Cille, near Crickhowell, and at Blaen Onnen, two miles to the west of Craig Cille, and is a small shrub clothing the cliffs up to 2000 feet altitude. The flowers and fruit are very similar to those of the mountain ash ; and Koehne supposes it to be a hybrid between *Pyrus intermedia* and *Pyrus Aucuparia*, which occur in the same locality.

The form *Mougeoti*, which is considered by many botanists to be a distinct species, occurs in Lorraine, the Vosges, Jura, Suabian and Western Alps, and in the Carpathians. It never attains a great size, being either a low bush or a small tree 15 to 30 feet in height. In Piercefield Park, Monmouth, Great Doward in Hereford, and a few other localities in the west of England, a shrub or small tree has been found which is near this form.[1]

Var. *scandica* has been found in Britain, in a few localities in Denbighshire and Breconshire,[2] and also at Chepstow[3] in Monmouthshire, always growing on limestone rocks. It was supposed to grow also in Arran, but Koehne,[4] as will be seen in our account of the peculiar forms of that island under *Pyrus pinnatifida*, denies its occurrence there.　　　　　　　　　　　　　　　　　　　　　　　　　　　(A. H.)

This variety is widely spread in Northern Europe. The best account we know of this tree is by Conwentz, who calls it "*Pirus Suecica*." He says that most authors speak of it as a small tree or shrub—Koehne only gives it as 7 metres high. It grows on granite, gneiss, chalk, and alluvium, and extends from the island of Åland, South-east Sweden, South Norway, and Denmark, to North-east Germany, where, however, it seems to be quite a rare tree and only recently discovered.

It is represented in France, Switzerland, Austria, and Bosnia by *P. Mougeoti*, which many botanical authors have mistaken for it, and which, according to Conwentz, can only be distinguished in some varieties by the fruit.

In the island of Oesel, in the Baltic, it is much planted, and often attains 2 metres in girth. Conwentz, however, found wild specimens at Soëginina near Karral, at Pajumois near Keilkond, at Wita Jahn, and in other places—mostly small trees, but

---

[1] It is called *Pyrus intermedia*, Ehrh., by the Rev. Augustin Ley. Briggs and Boswell think it is perhaps a form of *Aria* or *rupicola*. See *Jour. Bot.* 1884, p. 216. It is certainly quite distinct, in my opinion, from *scandica* or *latifolia*.

[2] *Jour. Bot.* 1903, p. 215.　　　　[3] Specimen at Kew.　　　　[4] *Jour. Bot.* 1897, p. 99.

in some places attaining 10 metres in height. In the Finnish islands of Åland it is found truly wild, in a few places only, sometimes in company with an allied species, *P. fennica.* Conwentz identified it at Bergö, Skarpnåtö, Labnäs, and elsewhere. The finest specimen he saw at Östergeta, being 12 metres high and 2 metres in girth.

In south-eastern Sweden it is more abundant, but does not occur in any of the provinces north of Wermland, about lat. 60° N. In the neighbourhood of Stockholm it grows at Stockby to 12 metres in height. In Södermanland and the island of Gothland it is more common.

In Denmark the tree has been found in many places, and is undoubtedly wild near Aarhus in Jutland, in the forests of Adslev, Kolden, and Jexen. I believe that I also saw it in the forest of Roldskov near Aalborg, though I did not at that time distinguish it from *Pyrus Aria.* In the island of Bornholm it is known under the name of "Axelbar."

In Germany it is confined to a limited area on the coasts of West Prussia and Pomerania, where Conwentz has found it living in six places only—Koliebken, Hoch Redlau, Oxhöft, Karthaus, Gr. Podel, and Markuhle near Kolberg. He gives maps showing the position of the trees in these places, and says that whilst *P. torminalis* grows in the interior, where the hornbeam is predominant, *P. intermedia* grows in the country along the coast, where the beech is the prevailing tree. It occurs most commonly in a shrubby condition, the tallest wild one being only 13 metres high by 1 metre in girth, but one tree at Gross Podel in Pomerania is 1.90 metre in girth, and at Wernigerode, in the Harz, a cultivated tree has attained 17 by 3.17 metres, which is the largest known to Conwentz. He thinks that the scarcity of the tree in Germany arises from its not being indigenous, as no geological evidence exists of its having been formerly commoner, and suggests that it has been introduced from Sweden by birds of passage, such as the waxwing or thrushes, which are fond of the fruit, and may have voided the seeds after migration from the north.

The Swedish name is *Oxel,* and this name being found in many place and family names in Sweden, shows that the tree was probably more common formerly than at present.

In Norway, Schübeler[1] says that it is wild only in the most southern parts, as at Porsgrund, Grimstad, and Dalen in Eidsborg, in lat. 59° 42′ N. There are large trees at Lunde in Stavanger district growing near the church. In the Botanic Gardens at Christiania I have seen a tree which is about 12 metres high and over 2 in girth. It has been planted and grows well at Stenkjær, at the north end of the Trondhjem Fjord. The Norsk name is *Maave.*

Dr. Brunchorst, Director of the Bergen Museum, informed the Earl of Ducie that *Pyrus intermedia,* as well as *P. pinnatifida* (*P. fennica*), were found on the south-west coast of Norway, and that a hybrid which he calls *Pyrus Meinickii,* *P. fennica* × *Aucuparia,* has also been recently discovered in the "Mosterö Bommel Fjord." Dr. Brunchorst, who has paid much attention to this genus, says that three species which he cultivates at Bergen vary much, and perhaps pass into one another.

---

[1] Schübeler, *Viridarium norvegicum,* vol. ii. 477 (1888).

I
Y

Lord Ducie has brought plants of these to Tortworth, where he grows them under the name of *Pyrus hybrida*.

REMARKABLE TREES

This species appears to be now rarely planted, except in botanical gardens. The best specimen which we have seen occurs at Syon (var. *scandica*). In 1904 it measured 48 feet in height by 7 feet 10 inches in girth, with a bole of 7 feet, dividing into 8 large branches, and forming a wide-spreading crown of foliage, about 50 yards in circumference (Plate 50). Another fine tree is growing at Livermere Park, Bury St. Edmunds, Suffolk, specimens and particulars of which have been kindly sent to us by Mr. Stiling. It is now (1905) 45 feet high by 8 feet 5 inches in girth, with a bole of 8 feet dividing into 12 main branches, the diameter of the spread of foliage being 45 feet. This tree was reported[1] in 1889 to have been 42 feet high by 8 feet 3 inches in girth. In August 1905 it was covered with fruit.   (H. J. E.)

There is a fine specimen at Stowe, near Buckingham, growing near the bridge over the lake in sandy soil, which measures about 45 feet high by 7 feet 9 in. in girth, with a 7 feet bole. It was loaded with fruit in August 1905.

At Wykeham Abbey, the Yorkshire seat of Viscount Downe, there is a fine tree on the lawn, about 40 feet high, spreading from the ground, where it measures 10 feet 8 inches in girth, into a large and well-shaped head.

This tree is planted in some of the parks and gardens in London, and grows well at the Botanic Gardens in Regent's Park. I am informed by Mr. A. Stratford, Superintendent of the Corporation Park of Blackburn, that it makes a good shade tree in that smoky town.

[1] *Garden*, 1889, xxxvi. 342. Note by J. C. Tallack, who named the tree *Pyrus pinnatifida*.

PYRUS PINNATIFIDA, Bastard Mountain Ash

Pyrus pinnatifida, Ehrhart, "Plantag." 22, ex *Beiträge zur Naturkunde*, vi. 93 (1791); Loudon,
   Arb. et Frut. Brit. ii. 915 (1838), N. E. Brown, in *Eng. Bot.* iii. ed. Suppl. 168 (1892); *Gard.
   Chron.* xx. 493, fig. 78 (1883).
Pyrus semipinnata, Roth, *En. Pl. Phæn. in Germ.* i. sect. post. 438 (1827).
Pyrus fennica, Babington, *Man. Eng. Bot.* ed. 3, p. 111 (1851).
Sorbus hybrida, Linnæus, *Sp. Pl.* 684 (1762); Schübeler, *Viridarium norvegicum*, ii. p. 476.
Sorbus fennica, Fries, *Summa Veg. Scand.* 42 (1846).

A species of hybrid origin, occurring as a small tree, which may attain 50 feet in height, with smooth, grey bark. Leaves variable in shape, mostly pinnate or deeply cut at the base, with 1-4 pairs of segments more or less separate; the upper part cut into deep sharp-toothed lobes; green and glabrous above, grey tomentose below. Flowers white in loose corymbs; styles 3, woolly at the base; fruit small, globular, coral red, and resembling that of *Pyrus Aucuparia*.

Varieties

This form, the parents of which are *P. Aucuparia* and *P. intermedia*, must be carefully distinguished (see p. 143) from *Pyrus hybrida*, Moench, a shrub of different origin.

Pyrus Thuringiaca, Ilse,[1] a cross between *P. Aucuparia* and *P. Aria*, is generally included under *P. pinnatifida*, from which it differs only in the leaf, whiter beneath, having its upper part lobulate or dentate and not deeply lobed.

Sorbus arranensis, Hedlund,[2] is the name given to a form occurring in the Isle of Arran, which is intermediate between *P. pinnatifida* and *P. intermedia*, and closely resembles the latter, differing only in the deeper and more irregular lobing of the leaf.

The hybrid forms, which are intermediate between *P. pinnatifida* and *P. Aucuparia*, are generally regarded as varieties (var. *satureifolia*[3] and var. *decurrens*[4]) of the latter species, and will be mentioned in our account of the mountain ash.

Identification

Pyrus pinnatifida and the intermediate hybrids are variable and inconstant in the shape of the leaf. There is no difficulty, however, in their identification, if it be noted that hybridity may be suspected in all cases where the leaves vary on the one hand from the regularly pinnate separate leaflets of *Pyrus Aucuparia*, and on the other from the regular uniform lobing or serration of *Pyrus intermedia* or *Pyrus*

---

[1] In *Jahresb. Bot. Gart. u. Mus. Berlin*, i. 232 (1881).
[2] In *Kon. Sv. Veten. Akad. Handl.* 1901-2, p. 60.
[3] Koch, *Dendrologie*, i. 189 (1869).
[4] Koehne, *Deutsche Dendrologie*, 248 (1893). This variety is commonly known as *Pyrus lanuginosa*, Hort.

Aria. In winter, specimens of cultivated *Sorbus fennica* show the following characters represented in Plate 45 :—

Twigs : long shoots glabrous, shining, dark brown, with a few scattered lenticels. Leaf-scar crescentic, very narrow, set obliquely on a reddish brown, slightly projecting cushion; it shows a varying number of bundle traces,[1] 3, 4, or 5, and may thus be distinguished from other species of Sorbus, as *Pyrus Aucuparia* has 5 dots on the scar, while *Pyrus Aria, intermedia,* and *latifolia* have only 3. Terminal bud large, conic, tomentose, especially at the apex. Lateral buds small, either appressed to the stem or diverging from it at an acute angle. Bud-scales few, densely pubescent on the outer surface, and ciliate in margin. Short shoots ringed, pubescent, bearing a terminal bud.[2]

DISTRIBUTION

The form *fennica* occurs plentifully in Scandinavia, where it grows wild, reproducing itself naturally by seed, and behaving as a true species. It extends in Norway, according to Schübeler, up to lat. 66° 14′ on the west coast as a wild plant, and in Sweden up to 60° wild and 62° planted; it also occurs in Finland, but is not recorded from other parts of Russian territory. In Central Europe it only occurs sporadically, and apparently always in company with the parent species; it is recorded from various mountain stations in France, Switzerland, Germany, and Austria.

The hybrids which occur in the Isle of Arran have attracted much attention and discussion. Formerly it was believed that *Pyrus Aucuparia, Pyrus intermedia* (var. *scandica*), and *Pyrus fennica,* all occurred in a wild state. Koehne,[3] however, considers that (excepting *Aucuparia*) all the plants in question on the island are hybrids, there being two sets, one typical *fennica,* while the other set comprises forms between that and *scandica.* This view, which excludes one of the parents (viz. *scandica*), implies that these hybrids, once established, may under favourable conditions reproduce themselves naturally and behave generally as true species.

N. E. Brown says of this species that it is "rare and perhaps not indigenous except in Scotland"; but he has seen specimens from Kent, Sussex, Hants, Somerset, Gloucester, Leicester, Stafford, Cumberland, Roxburgh, Arran, and Dumbarton. He thinks that Arran seems to be the only truly native locality for this tree in the British Isles, and believes that the Arran plant placed under *intermedia* is a form of it. Watson, however, states in his *Compendium,* p. 510, that Borrer held it to be wild in North Hants between Farnham and Farnborough, where it was observed sparingly along with *Aria* and *Aucuparia,* both more plentifully. A specimen picked by James M'Nab in Darenth Wood, Kent, is, according to Watson, identical with Arran specimens.

There is a fine tree on the edge of a shrubbery close to Wilton House, Wilts, the seat of the Earl of Pembroke, about 50 feet by 5.

---

[1] If the dots are not plainly visible externally, they can be seen clearly on paring off the epidermis of the scar.

[2] The twigs in winter described above clearly show the hybrid origin of this species; the varying number of dots on the scar, the pubescence and shape of the scales, etc. show the influence of *Pyrus Aucuparia.*

[3] Koehne, *Jour. of Bot.* 1897, p. 99. See also Rev. Dr. Landsborough's account of the Arran hybrids in *Trans. Bot. Soc. Edin.* xxi. 56 (1897).

There is a tree at Williamstrip Park, Fairford, the seat of Lord St. Aldwyn, which is in a decaying condition. It consists of a large stool measuring 8 feet at the ground, with four stems about a foot in diameter by about 60 feet high. There is also a tree at Arley Castle 50 feet high by 3 feet 5 inches in girth in 1905. One at Bayfordbury is 40 feet by 5 feet 2 inches, branching at 4 feet into four stems, with numerous ascending branches. At Aldenham Cottage, Letchmore Heath, Herts, is a fine tree 44 feet by 6 feet 2 inches, with a bole of 6 feet. At Danson Park, Welling, Kent, the residence of Mr. Bean, there is said to be a tree about 30 feet high, with a girth of 12 feet 4 inches at $1\frac{1}{2}$ feet above the ground, described to be like a large bush with seven main branches. (A. H.)

PYRUS ARIA, Whitebeam

Pyrus Aria, Ehrhart, *Beiträge zur Naturkunde,* iv. 20 (1789); Loudon, *Arb. et Frut. Brit.* ii.
    910 (1838).
Cratægus Aria, Linnæus, *Sp. Pl.* 475 (1753).
Sorbus Aria, Crantz, *Stirp. Aust.* ii. t. 2, f. 2 (1762).
Aria nivea, Host, *Fl. Aust.* ii. 8 (1813).

A tree in woods and on good soil attaining a height of 40 to 50 feet, and rarely
70 feet in height; but in rocky and mountainous situations usually remaining
shrubby.   Bark smooth and grey, becoming slightly fissured in old trees.   Leaves
stalked, oval or obovate, rounded, cordate, or cuneate at the base, sharp or obtuse
at the apex, biserrate or slightly lobulate with teeth, the lobules largest towards
the apex of the leaf; green and glabrous when adult above, but always snowy-
white tomentose beneath; nerves, 8 - 12 pairs, very prominent on both surfaces.
Flowers with an unpleasant odour, white, in loose corymbs; the peduncle,
receptacle, calyx, and corolla, white tomentose; styles 2, free, pubescent at the base.
Fruit globose or ovoid, $\frac{1}{2}$ inch in diameter, shining red with a few brown dots,
tomentose at the base and apex; flesh scanty, sweetish acid in flavour.

IDENTIFICATION

In summer the leaves, snowy white underneath and with prominent nerves, are
a sure guide.   The leaves of *Aria* from a wild specimen growing at Gosford, Kent
(Fig. 18); of var. *rupicola* from a wild specimen from north-west Lancashire (Fig.
13); of var. *salicifolia* (Fig. 16) and var. *Decaisneana* (Fig. 8), both from specimens
cultivated at Kew, are shown on Plates 43 and 44.   In winter the following
characters are available, as shown in Plate 45 :—

Twigs : long shoots round, shining brown, glabrous except for a little pubescence
near the tip, marked with scattered wart-like lenticels.   Leaf-scars set obliquely
on prominent leaf-cushions, crescentic, with three bundle traces.   Buds ovoid, conical-
pointed, shining, and somewhat viscid; terminal larger, side-buds coming off at an
acute angle.   Bud-scales glued together, strongly keeled, glabrous on the surface,
densely long ciliate in margin.   Short shoots ringed, generally glabrous, and
ending in a terminal bud.   Viscid buds occur also in *Pyrus Sorbus,* which is, however, very distinct in its five-dotted scars and glabrous scales.

VARIETIES

Some authors take *Pyrus 'Aria* in a wide sense, and under it group *Aria*
proper, *rupicola, latifolia, scandica,* etc. as sub-species.   Most of these, as being
readily distinguishable by many characters, have been considered by us as distinct
species.   Taking *Pyrus Aria* in a narrow sense, as comprising forms with leaves

snowy white beneath, it exhibits a great variety of forms in the wild state, explained by its wide geographical distribution and its occurrence on different soils and in different situations. Moreover, various horticultural varieties have been produced. The type has been described above; the following is a list of the most important varieties :—

1. Var. *rupicola*.[1] Differs from the type in the leaves having fewer nerves, generally 7 (5-9) pairs, less prominent; obovate-oblong in shape, widest above the middle, lobulate above, with the tomentum ultimately becoming slightly grey. Fruit smaller, ⅜ inch diameter, carmine-scarlet. This variety occurs always on limestone rocks, and is recorded from many stations in the British Isles. It is probably a form due to poor soil and exposure to wind, and other uncongenial conditions.

2. Var. *græca*, Boissier.[2] A shrub occurring in Spain, Albania, Greece, Syria, and Asia Minor. Leaves round, thick, almost leathery in consistence, nerves 6-10 pairs, broad, cuneate at the base, lobulate, with large teeth in the upper two-thirds. This form is also known as *Sorbus cretica*, Fritsch, and *Aria græca*, Decaisne.

3. Var. *flabellifolia*.[3] Leaves orbicular, cuneate, or rounder at the base, margin with large incisions, sharply toothed, nerves 3-5 pairs. South-eastern Europe and Asia Minor.

4. Var. *Decaisneana*, Rehder.[4] Leaves large, 4-6 inches long by 2-3½ broad, elliptic or oblong, narrow or acuminate at the apex, rounded or subcordate at the base, serrate in almost the whole margin with sharp teeth; nerves 12-14 pairs; petiole channelled above, nearly an inch long. Flowers first white, then becoming pinkish; styles glabrous. Fruit purplish, ellipsoid, crowned by the persistent hairy sepals. This tree is of unknown origin; it has been said to be Himalayan, but I am not aware on what authority. It closely resembles *Pyrus lanata* from that region.

5. Var. *sinensis*. Leaves narrow, lanceolate or ovate, with acuminate apex and cuneate base; crenately serrate. A series of forms occur in the mountains of Hupeh in China, where the trees are common at high elevations, and vary in size from 10 to 40 feet in height. Seeds were sent home by Wilson to Messrs. Veitch in 1901, and seedlings, very beautiful in foliage and vigorous in growth, are now growing at Coombe Wood.

6. Var. *salicifolia*. Leaves narrow, ovate-lanceolate, doubly serrate in margin, with long petioles. Origin unknown.

7. Certain horticultural varieties occur in which the leaves are variously coloured, as *lutescens, chrysophylla, sulphurea*.

8. Var. *quercoides*. Leaves regularly lobed with their edges bent upwards.

Pyrus Hostii,[5] Hort., may here be mentioned, as it occurs in cultivation and

[1] *Pyrus rupicola*, Boswell Syme, *Eng. Bot.* ed. 3, t. 483.
[2] *Flora Orientalis*, ii. 658. There is a form in south-east Europe called *meriodionalis*, which differs only slightly from this variety.
[3] *Cratægus flabellifolia*, Spach, *Hist. Vég. Phan.* ii. 103.
[4] Rehder in *Cyclop. Am. Hort.* iv. 1689 (1902). *Aria Decaisneana*, Lavallée, *Arbor. Segrez.* p. 51, t. 18. *Pyrus Decaisneana*, Nicholson, *Kew Hand-list of Trees and Shrubs*, 187 (1894). *Sorbus Decaisneana*, Zabel, *Handbuch Laubholz-Benennung*, 199 (1903).
[5] Figured in *Garden*, 1881, xx. 376.

resembles *Pyrus Aria*.   It is of hybrid origin, one parent being either that species or *Pyrus intermedia*, while the other is *Pyrus Chamæmespilus*.   It is distinguished from *Pyrus Aria* by the larger and more irregular teeth of the leaves (*cf.* Plate 44), and its flowers are pinkish white, borne in loose corymbs.   Various intermediate forms have been distinguished, as—

Sorbus ambigua, Michalet.   Exactly intermediate between *Pyrus Aria* and *Pyrus Chamæmespilus*, with the leaves larger than in the second, and smaller than in the first, and the margins having a tendency to lobing.   Tomentum whitish.

Sorbus arioides, Michalet.   A form intermediate between *ambigua* and *Aria*.

Chamæmespilus × *Mougeoti*.   Leaf large, with lobes well marked and rounded; tomentum greyish.   These hybrids are common in the Jura and the Alps.

DISTRIBUTION

The whitebeam is a wide-spread species.   It occurs throughout Europe generally, reaching in Norway as far north as lat. 63° 52', and in Sweden to lat. 59°.   It is met with also in Algeria, Asia Minor, the Caucasus, Armenia, Siberia, and Central China, assuming in some of these regions remarkable varietal forms.   It is replaced in the Himalayas and Japan by *Pyrus lanata*, Don, an allied species.

While it occurs on all soils except those which are wet, it has a decided preference for limestone.   In woods and hedges it grows to be a small tree; but in exposed situations on rocky mountains, etc. it dwindles to a mere bush.   On the Alps it ascends to 4800 feet.

In the British Isles its distribution has not been accurately made out, as many supposed records refer rather to *intermedia* or *latifolia*.   Apparently, however, as a wild tree, the typical form is almost entirely confined to the southern and midland counties of England and to south Wales.   Variety *rupicola* is recorded from nearly every county from Devon to Sutherland, and is widely spread on the littoral range between Lancaster and Humphrey Head, ascending in Banffshire to 1200 or 1400 feet, where it has been found by Dr. Shoolbred of Chepstow on limestone cliffs near Inchrory in upper Banffshire.   In Ireland the whitebeam is rare and local, and both the type and *rupicola* occur.   (A. H.)

REMARKABLE TREES

By far the finest specimen that we know of in England or elsewhere grows on the edge of Camp Wood, near Henley on Thames, on Sir Walter Phillimore's property, where Henry saw it in 1905.   It measures 75 feet high by 4 feet 9 inches in girth, with a bole about 35 feet long, and has very smooth beech-like bark (Plate 51).

There is a large and very well shaped tree at Walcot, Shropshire, which in 1906 Elwes found to be 56 feet high and 6½ feet in girth, with a clean bole about 20 feet long.

There is a handsome tree on the lawn at Belton Park, which measures 41 feet by 6 feet 7 inches.

A very spreading, ill-shaped tree in a thicket at Mount Meadow, near Cobham, Kent, is 9 feet 3 inches in girth.

At Stowe, near Buckingham, there are several fine trees near the Queen's Temple, which are about 50 feet high, but the tree when growing wild on the Cotswold Hills, where it is common, rarely exceeds 30 feet with a stem 2 to 3 feet in girth, and is more usually seen as a bush with many stems.

The whitebeam, like the mountain ash, is occasionally found as an epiphyte growing on other trees, where its seeds have been dropped by birds. Though this is more common in the damp climate of the west of England, yet we know of two cases which are remarkable on account of their situation. One is in the Yew Tree Vale in Surrey, where a whitebeam is growing near the top of a yew tree;[1] the other is near Colesborne in the Cotswold Hills. In this case a large limb has been torn by the wind from a Scots pine, and in the crevice on the east side of the tree, where but very little vegetable matter has yet had time to form, a healthy young whitebeam, now about 3 feet high, grew for seven or eight years, when it began to lose vigour.

Though it is well known that the decaying mossy trunk of a fallen tree is one of the most favourable situations for the seeds of many conifers to germinate and grow, yet in this case the roots of the whitebeam must derive their nourishment almost entirely from the air, the case being very different from those so often seen in the Himalayas and other countries, where a large quantity of moss, ferns, and decaying vegetable matter accumulate in the forks of large old trees.

The whitebeam is easily propagated by seed, which, if sown in autumn, will germinate partly in the following spring and partly in the second year after sowing. The seedlings grow slowly at first, and require five or six years in the nursery before they are large enough to plant out. When planted on good soil the whitebeam is a very ornamental tree, both on account of its leaves and fruit, which is larger and more abundant than when wild. It is, however, so much liked by birds that it is soon eaten up.

Timber

The wood is hard, heavy, and even in the grain, and is white in colour, with some dark spots, and in old trees becomes occasionally tinged with red. It is used on the Continent in turnery and in making tools.

Loudon says that it was used for the axletrees, naves, and felloes of wheels, carpenters' tools, and walking-sticks, but that the greatest use of its wood, until iron superseded it, was for the cogs of small wheels. I have felled a tree 18 inches in diameter, which when cut through was perfectly sound at heart, and was considered to be well suited for chair-making.

---

[1] *Garden*, 1882, xxii. 164.

I

Z

In Hampshire[1] the wood is used for making whip-handles, and the tree is known there on that account as "whip-crop."

Mr. Weale, of Liverpool, reports as follows on a sample of this wood which I sent him :—" The wood is of a medium hardness, good length of fibre, and takes a clean finish. Not tough. Rays invisible on transverse section. Grain moderately close and even. Warps badly in drying, and is liable to split."

In a little book on English timber by "Acorn" (Rider and Son, London, 1904), I find the following note on the whitebeam, though the author does not notice either of the service trees :—" In a green state whitebeam has a strong smell, and even after seasoning this is retained to a certain extent. A great many handles for cutlery are made from it, and its hardness is admirably adapted for these, as it is capable of taking a very high polish from the extreme closeness of its grain. It is also used in the manufacture of musical instruments, and the tops and small pieces are always appreciated by the turner." As it is usually available only in small pieces, these would probably, when thoroughly seasoned, be very useful to introduce as blocks in parquet flooring.   (H. J. E.)

[1] Townsend, *Flora of Hampshire*, 125 (1883).

TAXODIUM

Taxodium, Richard, *Ann. Mus. Par.* xvi. 298 (1810); Bentham et Hooker, *Gen. Pl.* iii. 429
    (1880); Masters, *Jour. Linn. Soc. (Bot.)* xxx. 24 (1893).
Schubertia, Mirbel, *Nouv. Bull. Soc. Philom.* iii. 123 (1812).
Glyptostrobus, Endlicher, *Syn. Conif.* 30 (1847).

DECIDUOUS or subevergreen trees, several extinct species and a series of living
forms, which have been variously considered to constitute one, two, or three species,
belonging to the tribe Taxodineæ of the order Coniferæ.

Branchlets of two kinds, those at the apex of the shoot persistent, and bearing
axillary buds, those lower down on the shoot deciduous and without buds. Buds of
two kinds : those near the apex of the shoot, two in number, sub-terminal, globose,
composed of imbricated, ovate, acute, keeled scales; these buds continue the growth
of the persistent shoot. The lateral buds, situated lower on the shoot, are minute
globose swellings, enclosed in two transverse, broadly oval, concave, membranous
scales, which do not meet. These buds produce the deciduous branchlets, and
are developed both on older and current year's shoots, in the latter case arising in
the axils of primary leaves.

Leaves inserted spirally on the branchlets; on the persistent shoots, spreading
more or less radially; on the deciduous shoots, in the usual forms of the species,
thrown by a twisting of their bases into two lateral ranks, thus assuming a pseudo-
distichous arrangement; linear, acute, channelled along the median line above,
keeled and bearing stomata below. In var. *imbricaria* the leaves are not pseudo-
distichously arranged, but are appressed around the twig and spreading at their
free apex; they are narrow, long-pointed, concave above.

Flowers monœcious. Male flowers in panicles, 3 to 5 inches long, arising at the
end of the preceding year's shoot. Each flower is minute, sub-sessile, and consists
of a stalk surrounded at its base by ovate scales, and bearing 6 to 8 distichously
opposite stamens. Female flowers, scattered near the ends of branchlets of the
preceding year, solitary, globular, consisting of numerous imbricated pointed bracts,
adnate below to the thickened fleshy scales, each of which bears two ovules.

Fruit, a globular or ellipsoidal, short-stalked, woody cone, an inch or more in
diameter, ripening in the first year, composed of thick coriaceous peltate scales, the
stipes of which are slender and spring off at right angles from the axis of the cone ;
the discs, rhomboidal in shape, show a triangular scar at the base, above which they
are irregularly crenulate and rugose. The bract having almost entirely coalesced

with the scale, its apex appears on the upper part of the scar as a minute reflexed point.   Some of the scales are sterile ; the others bear each two erect unequally three-angled seeds.

Taxodium is readily distinguishable in winter from other deciduous trees by the peculiar buds and branchlet scars which mark the twigs.   The latter are very slender, terete, glabrous, and brown in colour, and bear at their apex the two pseudo-terminal buds described above, one of which, however, is often aborted in trees growing in England.   Scattered over the twigs appear the branchlet scars and the lateral buds.   The former are small circular depressions, surrounded by a slightly raised rim, and having a single dot or a minute protuberance in their centre.   The lateral buds, also previously described, are smaller than the branchlet scars, and on twigs of one year arise just above the minute scars left by the primary leaves, in which a single dot may be made out with difficulty.   Single-dotted leaf-scars occur in Larix and Pseudolarix ; but in these genera branchlet scars are absent, and the twigs show spurs or short shoots, which are wanting in Taxodium.

The genus Taxodium was once common and widely distributed over the Holarctic region.   During Miocene and Pliocene times it was spread over the interior of North America, throughout Europe, and in north-eastern Siberia.   In the present day it is restricted to the Southern United States and Mexico.

The genus can only be confounded with *Glyptostrobus*, now represented by one living species, *G. heterophyllus*, Endlicher,[1] a native of the province of Canton, in Southern China, where it occurs· as a small tree along the banks of rivers and streams.   Like Taxodium, it has deciduous foliage and branchlets.   The leaves assume two forms—on ordinary branchlets long and linear and arranged in three rows, on fruiting branchlets closely imbricated, scale-like, concave internally and carinate externally.   The cone, pyriform in shape, is composed of scales, which are not peltate, but elongated and arising from its base.   The bract coalesces with the scale below ; but above the middle is free and recurved, leaving bare the 5 to 7 lobed summit of the scale.   The seeds, oblong or obovate, often short-spurred at the base, are narrowly winged on the sides and prolonged at the base into a flat, lancet-shaped wing.   *Glyptostrobus heterophyllus* is not hardy at Kew, where specimens may be seen in the temperate house.   A plant of it is reported to be growing in the open air at Castlewellan.

[1] *Glyptostrobus heterophyllus*, Endlicher, *Syn. Conif.* 70 (1847) ; Masters, *Jour. Bot.* 1900, p. 37, and *Gard. Chron.* xxvi. 489 (1899) ; *Thuya pensilis*, Staunton, *Embassy to China*, ii. 436 (1798) ; Lambert, *Pinus*, ed. 2, ii. 115, f. 51.

TAXODIUM DISTICHUM, Deciduous Cypress

Taxodium distichum, Richard, *Ann. Mus. Par.* xvi. 298 (1810); Loudon, *Arb. et Frut. Brit.* iv. 2481 (1838); Sargent, *Silva N. America*, x. 151, t. 537 (1896); Kent, in Veitch's *Man. Coniferæ*, 281 (1900).

Cupressus disticha, Linnæus, *Sp. Pl.* 1003 (1753).

Schubertia disticha, Mirbel, *Mém. Mus. Par.* xiii. 75 (1825).

Three well-marked forms of Taxodium occur in the wild state, which differ in certain characters, such as the form of the foliage, its partial persistence or complete deciduousness, and the time of flowering; and in the present state of our knowledge these may be considered as constituting one species, the peculiarities mentioned appearing to depend on conditions of soil and climate, and to be by no means constant.

1. Var. *typica*. A tall tree, with a gradually tapering stem, which has an enlarged base, usually hollow internally and buttressed externally. When young it is strictly pyramidal in form; but in older trees the crown becomes wide and flattened, often 100 feet across, according to Sargent. The bark is dull reddish brown, 1 to 2 inches thick, fissured and separating into long fibrous scales. The leaves in this form are arranged pseudo-distichously on horizontally spreading branchlets, and are linear in shape (see generic description). This form is the one which occurs generally in the alluvial swamps of the south-eastern United States.

2. Var. *imbricaria*.

> *Taxodium distichum*, var. *imbricaria*, Sargent, *l.c.* 152.
> *Taxodium distichum pendulum*, Carrière, *Conif.* 182 (1867).
> *Taxodium imbricarium*, Harper, *Bull. Torrey Bot. Club*, xxix. 383 (1902), and xxxii. 105 (1905).
> *Taxodium sinense*, Gordon, *Pinetum*, 309 (1858).
> *Cupressus disticha*, β *imbricaria*, Nuttall, *Gen.* ii. 224 (1818).
> *Glyptostrobus pendulus*, Endlicher, *Syn. Conif.* 71 (1847); Hooker fil., *Bot. Mag.* t. 5603 (1886).

A tree, generally smaller in size than the type, with branchlets normally erect, but occasionally somewhat spreading and very rarely pendulous. Leaves appressed on the branchlets and acicular-acuminate (see generic description).

According to Mohr,[1] this is the "upland Cypress" which occurs on the shallow ponds of the pine-barrens and in semi-swampy woods on poor sandy soil. He considers it to be greatly inferior to the typical cypress of the alluvial swamps in regard to the size and quality of the wood; and states that in the earlier stages of its growth and on vigorous adventitious shoots it produces leaves of the ordinary form. It passes readily, according to his observations, into the type, where the soil conditions are favourable. He considers the peculiarity of the foliage to be an adaptation to check excessive transpiration during the time of drought when the sandy soil is laid bare to the sun and the supply of water diminishes.

---

[1] *Contrib. U.S. Nat. Herbarium*, vi. 117 and 325 (1901).

Harper considers this variety to be a distinct species, and in support of this opinion alleges that certain differences which he has observed in the two forms are constant.   The bark in var. *imbricaria*, both in cultivated and wild specimens, is considerably thicker and more coarsely ridged than in the typical form.   The enlargement of the base of the trunk is abrupt in the former, conical in the latter. Knees are formed more abundantly in trees of the type, and are usually slender and acute, sometimes reaching a height of 6 feet.   In var. *imbricaria* the knees are short and rounded, often almost hemispherical in shape.   The type is a lover of limestone, the variety just the opposite.   The distribution of the two forms is different, dependent upon the geological nature of the soil, var. *imbricaria* always growing on the Lafayette formation, which is a deposit of sandy clay, while the type always occurs on other formations.   Harper admits the occurrence of intermediate forms, but states that they are rare.   He has records of 300 to 400 stations in Georgia for var. *imbricaria*, at each of which there may be from ten to several thousand trees, while he has only seen intermediate forms about twenty times, and never more than 100 trees at one station.   In the intermediate forms branchlets with distichous leaves occur on young shoots.   Harper has seen in Georgia specimens of var. *imbricaria* as large as the ordinary form ; but it is generally admitted to be a smaller tree.   The two forms often grow close together, but in different situations.   On the Savilla river in Camden County, Georgia, he noticed the type growing along the water's edge below the Lafayette formation, while a hundred yards or so away var. *imbricaria* was flourishing in moist pine-barrens.

Var. *imbricaria* is possibly a juvenile form, analogous to *Cryptomeria elegans*. The generally smaller size of the trees and the various differences noted by Harper are probably the result of poor soil, and do not, in my opinion, entitle this form to rank as a distinct species.

This variety was early introduced into England, as it was in cultivation, according to Aiton,[1] at Kew in 1789.   The original tree at Kew, now dead, was living in 1886, when it was described by Sir Joseph Hooker[2] as 40 feet in height and of remarkable habit, on account of its slender twisted stem with decurved branches and pectinately-disposed branchlets.   A small tree, 20 feet in height, is now growing in Kew Gardens.

A tree of the Mexican kind was reported[3] to be growing at Penrhyn Castle, North Wales ; but Elwes saw it in 1906, and confirms the opinion I had formed from specimens sent by Mr. Richards, that it is var. *imbricaria*.   It is 44 feet high and 4 in girth, and comes into leaf later than the ordinary form growing near it.

At Pencarrow,[4] Cornwall, there is a fine specimen, which was planted about 1841 by Sir W. Molesworth.   It had attained in 1899 a height of over 30 feet, with a girth of stem of 2 feet 9½ inches at 5 feet from the ground.

---

[1] *Hortus Kewensis*, iii. 372.   Described as "Cupressus disticha, var. nutans ; foliis remotioribus subsparsis : long-leaved deciduous cypress."   This varietal name was kept up by Loudon, *loc. cit.* 2481, who considered it to be identical with the *Taxodium sinense* of cultivators of his time.

[2] *Bot. Mag.* t. 5603 (1886), where it is described as *Glyptostrobus pendulus*, Endlicher.

[3] A. D. Webster, *Hardy Coniferous Trees*, 115 (1896).   This tree is described in *Garden*, 1887, xxxi. 480.

[4] Figured in *Gard. Chron.* 1899, xxvi. 489, fig. 161.

As ordinarily seen in cultivation it is a small tree of slow growth, and is quite distinct from the Chinese *Glyptostrobus heterophyllus*, with which it has been occasionally confused.

3. Var. *mucronatum*.

> *Taxodium mucronatum*, Tenore, *Ann. Sc. Nat.* sér. 3, xix. 355 (1853).
> *Taxodium mucronulatum*, Sargent, *Silva N. Am.* x. 150, note 2 (1896).
> *Taxodium Montezumæ*, Decaisne, *Bull. Soc. Bot. de France*, i. 71 (1854).
> *Taxodium mexicanum*, Carrière, *Traité Conif.* 147 (1855).
> *Taxodium distichum mexicanum*, Gordon, *Pinetum*, 307 (1858).

This differs from the type in the foliage being more persistent, generally lasting two years on the tree, and in the time of flowering, which is in autumn. The panicles of male flowers are generally more elongated than those of the United States tree. The leaves are usually shorter, lighter green in colour, and blunter at the apex.

These differences scarcely entitle this form, which occurs in Mexico, to separate specific rank. Specimens [1] of the type, occurring at high elevations (1600 to 2000 feet) in Texas, approach it in character of the foliage; and in some Florida specimens the panicles of flowers are as large as any occurring on Mexican trees. The cones vary greatly in size and form in trees of Taxodium, occurring both in Mexico and the United States. Sargent, who has seen the tree in Mexico, was unable to distinguish it, by either foliage or habit, from the type.

It is evidently a geographical form in which certain differences of foliage have been brought about by climatic influence. One is led by a study of the specimens from many different regions to see in Taxodium a single species very variable in the wild state, rather than a number of distinct species.

Taxodium does not produce knees, so far as we can learn, in Mexico, where trees generally stand upon dry ground. According to Seeman,[2] the tree is known in Mexico as *Sabino*, and is diffused over the whole tableland of that country. There are reported to be extensive forests of it at altitudes varying from 4500 to 7500 feet. Concerning, however, the character and distribution of these forests our information is very scanty. Much more is known about the remarkable isolated examples of very old and enormous trees, which have always attracted the attention of travellers in Mexico. The most noted of these is the tree of Santa Maria del Tule, about eighteen miles south-east of the city of Oaxaca, which was measured by Baron Thielmann [3] in 1886, when its height was between 160 and 170 feet. Its

---

[1] Specimens collected by Hillier in Keir County, Texas, are in the Kew Herbarium.

[2] *Botany of Voyage of H.M.S. "Herald"* (1852-1857), p. 335.

[3] *Garden and Forest*, 1897, p. 123; figured on p. 125. The tree is also depicted in *Gard. Chron.* 1892, xii. 646, fig. 100. According to a correspondent, the girth was 139 feet in 1886; 25 years previously it had been 136½ feet. Various and conflicting measurements of this tree, taken by Exter, Baron von Karwinski, and Galeotti, in the early part of the nineteenth century, are given by Zuccarini in *Ray Society, Reports on Botany* (1846), p. 19. The latest measurements of this tree I know of are on a very fine photograph given me by the late Hon. Charles Ellis, as follows :—

TAXODIUM DISTICHUM AT MITLA, NEAR OAXACA.—Reported dimensions—

Girth at 4 feet from ground, 132 feet.
,, ,, 6 ,, ,, ,, 154 ,,
,, higher up . . 198 ,,
Height, 100 to 120 feet.

(H. J. E.)

actual girth at 5 feet from the ground, following all sinuosities, was 146 feet, the longest diameter being 42 feet. The cypress of Montezuma, which is the largest of the great trees in the gardens of Chapultepec, near Mexico, is about 48 feet in girth, according to Elwes, who saw it in 1888. Its height is about 170 feet.[1]

Taxodium mucronatum was first described[2] from a specimen growing in the Botanic Garden at Naples, said to have been introduced into Europe in 1838. Elwes saw this tree in April 1903, when the old leaves were partly persistent. A tree at Palermo has borne fruit. There are specimens at Kew labelled "Hort. Cusinati," collected by J. Ball, which bear very large cones, 1½ inches long by an inch in breadth.

Two seedlings were raised by Elwes from seeds brought by Mr. Marlborough Pryor from Oaxaca in 1904, one of which is to be planted out in a sheltered dell at Tregothnan in Cornwall, the other in the Temperate House at Kew. The larger of these, which grew slowly in a greenhouse through the winter of 1904-5, was about 18 inches high at one year old.

The typical form is the one commonly cultivated in England. In summer the foliage is decidedly ornamental, being of a delicate green colour. In autumn the leaves, before they fall, become reddish brown in colour.

Sub-varieties.—About a dozen sub-varieties are enumerated by Beissner,[3] pyramidal, pendulous, fastigiate, dwarf forms, etc. The tree is very variable in habit.

Taxodium distichum rarely produces flowers or fruit in England. It first bore fruit about the year 1752. A tree[4] at Ryton-on-Dunsmore, which was forty years old, produced flowers, apparently all males and in great abundance, in 1868. Fruiting specimens were sent to Dr. Masters[5] from Menabilly in Cornwall in 1893; the cones were smaller than native-grown ones. One of these was proliferous, the cone terminating in a branch bearing leaves and male flowers; and from the sides of the cone leaf-bearing branches also emerged, which on examination proved to form no part of either bract or scale, but were separate outgrowths from the axis of the cone. On a tree at Gwydyr Castle, North Wales, fruit is borne about every third year, but Mr. Macintyre informed me that it never was fully matured, and no seedlings were ever raised. According to Webster,[6] this tree was profusely covered with cones in 1884, but had none when Elwes saw it in 1906. Bunbury[7] states that at Abergwynant, in Wales, a tree produced oval cones.

Gay[8] says that though often cultivated in wet places in several old parks at Paris, he has only seen fruit at the Trianon on a tree growing in very dry ground.

[1] *Garden and Forest*, 1890, p. 150, fig. 28.   [2] Carrière, *Traité Conif.* 147 (1855).
[3] *Nadelholzkunde*, 152 (1891).   [4] *Gard. Chron.* 1868, p. 1016.
[5] *Ibid.* 1893, xiv. 659, fig. 105, showing fruiting branch, scales, and seeds. In the same journal, 1886, xxvi. 148, fig. 28, are represented abnormal flowers of this species, from a tree growing in England; also, in *Gard. Chron.* 1888, iii. 565, fig. 77, is depicted a remarkable gnaur on a Taxodium.
[6] *Woods and Forests*, 1885, p. 25.   [7] *Arboretum Notes*, 161.   [8] Note in Kew Herbarium.

Seedling.—There are 5 or 6 cotyledons, borne in a whorl at the summit of a purplish brown caulicle, about 2 inches long, ending in a tiny curved rootlet, which subsequently develops a few lateral fibres. The cotyledons are linear, 1 to $1\frac{1}{4}$ inch long, $\frac{1}{10}$ inch broad, sessile on a broad base, gradually diminishing to an acute apex, upper surface dark green, bearing stomata in lines with a raised midrib; lower surface pale green and uniform. On the stem above the cotyledons are borne about 3 false whorls of leaves, $\frac{1}{2}$ inch long, those below resembling the cotyledons, but bearing stomata on both surfaces; those above having decurrent bases. In the axils of the uppermost leaves lateral branchlets are given off, bearing needles in two rows and forming short shoots, which fall off in autumn.

The preceding description is taken from seedlings raised at Colesborne from cones gathered by Elwes in September 1904 at Mt. Carmel, Illinois.   (A. H.)

DISTRIBUTION

This remarkable tree occurs in North America from southern Delaware, where, according to Sargent, it formerly attained almost its largest size, all along the coast region as far as the Devil's River in Texas, and up the Mississippi valley as far as southern Illinois and south-western Indiana. In these regions it inhabits river bottoms usually submerged during several months, and swampy places. On the Edwards Plateau of Texas,[1] several hundred miles west of the great cypress swamps of eastern Texas, it occurs at 1000 to 1750 feet above sea-level, and attains an enormous size at the edges of the deeper holes near the heads of the permanent water of the Pedernales and other streams. This highland form in certain respects resembles the Mexican variety. In some parts of Louisiana, Texas, and the Gulf States, it occurs as pure forest, and in places so continuously flooded that the seed cannot germinate. I have passed on the railway, built on trestles for miles, through cypress swamps where the soil was submerged to a depth of 5 or 6 feet, and where few other trees could live. In drier places, such as the Wabash valley in southern Indiana, near Mount Carmel, where the cypress is evidently not so happy, it was associated with ash, liquidambar, and maple. In this locality also, although the trees were covered with fruit, I could find no seedlings; and as the accessible trees are in most places being rapidly cut for their timber, they seem likely· to become scarcer unless protected. As far as I know it does not grow from the stool or from suckers.[2]

In Arkansas and Missouri there are swamps[3] in which both Taxodium and *Nyssa uniflora* grow together, the latter with a peculiar dome - shaped base,

---

[1] *Ann. Report U.S. Geol. Survey*, xviii. 210, 211 (1898). There are specimens from this locality at Kew.

[2] R. Ridgway describes this locality as being in 1873 heavily timbered with cypress over an area of about 20,000 acres, in which the best trees had even then been cut and floated out into the river. The largest stump he measured was 38 feet in girth at the ground and 22 feet at 8 feet high. The largest standing tree measured was 27 feet in girth above the swollen base, and the tallest 146 and 147 feet high. Their average height, however, was not above 100 feet, and even the finest of them would not compare for symmetry and length with the sweet gums (liquidambar) and ashes (*Fraxinus americana*) with which they were associated.—*Proc. U.S. Nat. Mus.* 1882, p. 87. An excellent photograph taken here is published in *Garden and Forest*, iii. p. 7, and shows the knees remarkably well.

[3] Coulter, *Missouri Bot. Garden Report*, 1903, p. 58.

analogous to the cone-shaped base of the former; and from Coulter's observations it would appear that seedlings of Taxodium are also rare here, and that it is being beaten in the struggle by the Nyssa, the seedlings of which are very abundant.

A disease[1] due to a fungus has attacked many of the trees in the Mississippi valley; the heartwood is found when the trees are cut down to be full of holes $\frac{1}{4}$ to $\frac{3}{4}$ inch in diameter.

Taxodium is one of the most striking and characteristic trees in the Gulf States, having its branches often covered with *Tillandsia usneoides*, the " Spanish moss " of the inhabitants, the long grey masses of which wave in the wind and give it a strange appearance.   The trunk takes many curious forms, which seem to be induced by the nature of the soil and the depth of the water, sometimes branching low and surrounded by buttresses, sometimes growing straight up to a considerable height (Plates 52-53).[2]   From the stout wide-spreading roots arise woody cylindrical projections, sometimes above a foot in diameter and 5 to 7 feet high, which are called " cypress knees."   The growth and functions of these have been the source of much discussion.[3]   Berkeley[4] supposed that they serve to aerate the submerged roots; others have thought that they help to anchor the roots in soft muddy soil. As the knees, however, occur to some extent even on ground which is never flooded, as in the trees at Syon, these suppositions, though highly probable, must remain somewhat doubtful.

The knees are hollow inside, and smooth externally, being covered with a reddish, soft, and spongy bark.   They never show any sign of vegetation, and will not put forth shoots, even if wounded and covered with earth.

CULTIVATION

In England the Taxodium grows much better than might be expected considering how much colder and shorter are our summers than those of its native country. It was introduced by John Tradescant about 1640, and described by Parkinson[5] as *Cupressus americana.*

For some unexplained reason it has lost the popularity it once enjoyed, and is now seldom planted, though it grows well in the southern and western counties.   I have raised it from American seed, which, however, must be soaked in warm water for some time, and placed in a warm house to get good results.   It grows rapidly at first, but as the young wood is not ripened, and no terminal bud formed (which

[1] Coulter, *Missouri Bot. Garden Report*, 1899, p. 23.

[2] For the negatives of the first of these photographs I am indebted to Miss E. Cummings of Brookline, Mass., a lady who is second to none in her love of and knowledge of trees.   The second, which was sent by Mr. W. Ashe, represents a typical cypress swamp on the Roanoak river, North Carolina, which has never been cut for timber.

[3] Sargent, *loc. cit.* 152, note 1 ; Coulter, *loc. cit.*   The best review I know of the literature on this subject is in a letter by R. H. Lamborn in *Garden and Forest*, iii. p. 21, which should be consulted by those interested, and which is illustrated by a very curious photograph, taken at Lake Monroe in Central Florida, of the denuded roots of the tree, showing that in some cases, at least, the anchor theory is proved.

[4] *Gard. Chron.* 1857, p. 549.

[5] Parkinson, *Theatr.* 1477, fig.   In *Catalogue of Trees*, London, 1730, p. 25, it is stated that the first tree, raised in Tradescant's garden near Lambeth, was then still living, being 40 feet high by 2 fathoms in girth.

Sargent says is also the case in America), the young plant must be kept under glass for the first two or three winters in order to develop a straight leader.

Many of the old trees which are to be found in England have evidently suffered from spring and autumn frosts when young, and have become stunted in consequence, but when the wood is ripe the tree will stand as much as 30° to 40° of frost, and I have seen it existing in the open air as far north as Copenhagen.

It should be planted in deep, moist loam, and the most sheltered situation that can be found, and may then be expected in the south and south-west to grow into a very fine and ornamental tree.

REMARKABLE TREES

The trees at Syon have been frequently described and figured. They are planted in damp soil by the side of a sheet of water, and one of them has produced knees of 1 to 2 feet high. This tree, which is shown in Plate 54, measured, in 1903, 90 feet by 12, but there is a much taller one on the other side of the water, which, when we saw it last in 1905, was 110 feet high, and is the tallest we know of in Europe. Another in the Duke's walk is 85 feet by 10 feet 3 inches.

But those at Whitton, near Hounslow, are even more remarkable, and are believed to have been planted by the Duke of Argyll between 1720 and 1762. They grow on gravelly soil, which, though apparently dry, is probably underlaid by damp alluvium. There are five trees standing in a group, of which the largest, carefully measured by us both in 1905, was 98 to 100 feet high by 13 feet 6 inches in girth; the others are all large, healthy, and growing trees (Plate 55).

At Pain's Hill, Surrey, there are two good trees: one,[1] measured by Henry in 1904, is 90 feet by 10 feet 9 inches, the other is 80 feet by 8 feet 6 inches.

At Parkside Gardens, Wimbledon, a tree is growing which is remarkably like the Ginkgo at Kew in habit. The bole at 7 feet divides into two stems, which give off seven or eight ascending branches. In 1904, measured by Henry, it was 65 feet by 11 feet 2 inches.

At Gothic Lodge, Wimbledon, the residence of Sir William Preece, there is a tree with a fine bole of 20 feet, dividing into several upright stems. In 1904, measured by Henry, it was 90 feet by 11 feet. This is perhaps the tree mentioned by Miller,[2] who says that a "tree at Wimbledon in the garden of Sir A. Janssen, Bart., bore cones for some years past and seeds which have been as good as those brought from America."

At White Knights, Reading, there are several trees, but none of large size, the biggest measuring, in 1904, 67 feet by 7 feet 10 inches. They are remarkable, however, for variety of habit. One is a tall, narrow tree with upright branches, almost fastigiate. In another tree the stem is twisted, as often occurs in the chestnut, and most of the branches are twisted also in the direction against the sun. Loudon mentions these as young trees of peculiar habits.

[1] This is probably the tree, reported in *Woods and Forests*, February 4, 1885, to be 83 feet in height by 10 feet in girth at 3 feet above the ground.

[2] Miller, *Gard. Dict.*, ed. 8, sub *Cupressus disticha* (1768).

At Barton, Suffolk, there are three trees, which measured in 1903, (*a*) in the Arboretum, 50 feet by 5 feet 5 inches, dying; (*b*) a smaller tree beside it, in a worse condition; (*c*) on the lawn, 56 feet by 4 feet 3 inches. The latter tree[1] was planted in 1826, the other two in 1831. It is evident that the dry though deep soil at Barton is not favourable to the growth of this species.

At Frogmore, Windsor, there are two specimens very different in habit. One, a clean-stemmed tree, growing near water, but without knees, is 80 feet by 8 feet 6 inches. The other, not so large, has a weeping habit, and is branched to the ground.

At Strathfieldsaye there is a tree, mentioned by Loudon as being 46 feet in height by 3 feet 4 inches in diameter, which I found in 1903 to be 63 feet high by 9 feet in girth. It is growing in stiff clay soil and has no knees; the stem is deeply furrowed.

At Dropmore there is a tree beside a pond, planted in 1843, and now measuring 60 feet by 5 feet 9 inches.

At South Lodge, Enfield, a tree is growing near water, with small knees, which, measured by Henry in 1904, was 77 feet by 11 feet 10 inches.

At Combe Abbey, Warwickshire, Mr. W. Miller[2] reports that a tree, mentioned by Loudon as 47 feet by 2 feet 3 inches in 1843, had attained, in 1887, 75 feet by 11 feet 6 inches at 3 feet from the ground.

At Longford Castle,[3] Salisbury, there are two trees, growing within a few yards of the river Avon. One, very tall, has a straight trunk free from branches for about 30 feet, and a girth of 8 feet 10 inches at 4 feet from the ground. The other is 6 feet in girth, and branches at 7 feet up.

At Brockett's Park, near Hatfield, the residence of Lord Mountstephen, there are many trees planted along a walk on the banks of the Lee, and forming an irregular line in which the trees vary very much in size. In the sheltered part of the valley, where the soil and situation are very favourable, they average 70 to 80 feet high, the best I measured being 80 feet by 10 feet and 86 feet by 9 feet. But lower down the stream, where the valley is more exposed to the wind, they are stunted, and not more than half the height of those above. There are knees on some of the trees overgrown with moss and meadowsweet, but not so large as those at Syon.

At Upper Nutwell, near Exeter, there is a tree which Mr. G. H. Hodgkinson informed me in June 1904 was 84 feet high by 11 feet 9 inches in girth.

Large trees have been reported at many other places, especially in the south of England, viz. :—

Connington Castle,[4] Huntingdonshire, a tree 70 feet by 7 feet in 1877; Watford,[5] Herts, 85 feet by 14 feet in 1884; Stanwell,[6] Surrey, a tree 13 feet in girth in 1904; Embley,[7] near Romsey, Hampshire, a tree $8\frac{1}{2}$ feet in girth in 1872, standing on the top of a hill.

[1] Bunbury, *Arboretum Notes*, 161.   [2] *Gard. Chron.* 1905, xxxvii. 12.   [3] *Garden*, 1890, xxxvii. 538.
[4] *Ibid.* 1877, xii. 405.   [5] *Woods and Forests*, 1884, p. 546.   [6] Reported by Sir Hugh Beevor.
[7] Bunbury, *Arboretum Notes*, 161.

I have seen no trees in Scotland of any size, and Henry has heard of none in Ireland, but there is one in the Edinburgh Botanic Gardens 31 feet by 3 feet in 1905.

TIMBER

According to Sargent the timber is light and soft, close, straight grained, not strong, easily worked, and very durable in contact with the soil. It is largely used for building, most of the houses in Louisiana and the Gulf States being built from it, and large quantities are also now exported to the North, where it is found a most valuable wood for doors, sashes, balustrades, and greenhouses.

The Stearns Lumber Company of Boston, U.S.A., are making a speciality of it, and from a pamphlet published by this firm I take the following particulars :—

The timber varies considerably in different localities, and they consider, after long experience, that the so-called Gulf Cypress, grown in Florida, is better than the Louisiana Red Cypress, or that from the Atlantic coast of Georgia. Farther north it is apt to be more shaky and of coarser grain ; and it is claimed that the seasoning is better done in the South than in the Northern States, from one to five years being required to do this properly, according to the dimensions of the timber, and that the longer in reason that it is kept in the pile before using the better.

It is said to be more durable, and to shrink and swell less than spruce or pine, to take paint well, and, as it contains no pitch, to resist fire longer than other coniferous woods.

It is quoted from the *Richmond Despatch* that a house, built by Michael Braun in 1776, and still owned and occupied by his descendants, was covered with cypress shingles, which were only removed in 1880.

Such shingles are now made by machinery at a very low price, and would be well worth trying for roofing houses in England, as they are very light in weight and inexpensive, and though I have no evidence that they are better than shingles made from English oak, their much greater size makes them easier to lay, and they can be cut to fancy patterns, which makes them very ornamental for roofing.

This wood is also highly recommended for doors, sashes, tanks, and other purposes where a great power of enduring damp is required.

It occasionally produces very ornamental wood, which is mottled and grained with red and brown, and some doors made of this wood, two of which I now possess, are extremely handsome.

Whether the wood grown in England will prove equally good I cannot say, as large trees are so seldom cut down in England that I have been unable to try it, but would certainly advise anyone who may have the opportunity to do so.

(H. J. E.)

THUYA

Thuya,[1] Linnæus, *Gen. Pl.* 378 (1737); R. Brown, *Trans. Edin. Bot. Soc.* ix. 358 (1868); Bentham et Hooker, *Gen. Pl.* iii. 426 (*ex parte*) (1880); Masters, *Jour. Linn. Soc. (Bot.)* xxx. 19 (1893). *Biota*, Endlicher, *Syn. Conif.* 47 (1847).

EVERGREEN trees of pyramidal habit and aromatic odour, belonging to the tribe Cupressineæ of the order Coniferæ. Branches spreading and much ramified terminating in so-called "branch-systems," which are flattened in one plane and are 2-, 3-, or 4- pinnately divided, their primary and other axes being densely clothed with scale-like leaves. These branch-systems[2] when they fall are cast off as a whole, the leaves not falling separately. The leaves, which are minute, are more or less coalesced with the axes, on which they stand in 4 ranks in 2 decussate pairs, those of the lateral ranks being conduplicate or boat-shaped, those placed dorsally and ventrally being flattened. In the seedling stage and certain horticultural varieties,[3] the foliage is different, the leaves being acicular, spreading, and uniform; all 4 ranks in this case are alike.

Flowers monœcious, all solitary and terminal on the ultimate short branchlets of the preceding year, the male and female flowers on different branchlets, the former on the branchlets near the base of the shoot, the latter on those near its summit. Male flowers cylindrical or globular, consisting of 3 to 6 pairs of stamens placed decussately on an axis, each with an orbicular connective bearing 2 to 4 pollen sacs. Female flowers minute cones, composed of opposite scales in which no distinction of ovular scale and bract is visible, continuous in series with the leaves at the end of the branchlet, 2 to 4 pairs in *Biota*, 4 to 6 pairs in *Euthuya*, mucronate at the apex, some sterile, the others fertile and bearing 2 to 3 ovules.

Cones solitary, ultimately deflected, except in *Biota*, in which they retain the erect position, oblong, ovoid, or almost globose, composed of 3 to 6 pairs of decussate scales, which are not peltate, some fertile, the others sterile, the uppermost often united together. Seeds 2 to 3 on each fertile scale. Cotyledons 2.

The genus Thuya, as understood here, does not include Chamæcyparis and

---

[1] *Thuya* has been written *Thuja* in Linnæus, *Hort. Cliff.* 449 (1737), and *Sp. Pl.* 1002 (1753); and *Thuia* in Scopoli, *Introd.* 353 (1777).

[2] The branchlets become brown in colour before they fall. See Masters, *Gard. Chron.* 1883, xx. 596.

[3] In addition to the varieties, in which the foliage retains permanently the seedling character, other forms occur in cultivation, in which the leaves are intermediate in shape between those of the seedling and of the adult plant. These varieties resemble the so-called *Retinospora* forms of the genus Cupressus, and were formerly considered, like them, to belong to a distinct genus.

Thujopsis, which were united with it by Bentham and Hooker. So limited, it comprises 5 species, and is divided into the two following sections :—

I. *Euthuya.* Cones with thin, coriaceous mucronulate scales, those of the 2 or 3 middle ranks being fertile. Seed thin, with lateral wings and a minute hilum. This section comprises 4 species, *Thuya occidentalis* and *Thuya plicata* of North America, *Thuya sutchuenensis* of central China, and *Thuya japonica* of Japan.

II. *Biota.* Cones with thickened, conspicuously umbonate scales, which are fleshy when young, almost ligneous when ripe; those of the lowest two ranks fertile. Seed thick, without wings, the hilum being large and oblong. This section includes one species, *Thuya orientalis* of north China.

The Thuyas resemble considerably in foliage and habit the flat-leaved cypresses. The latter are best distinguished by their fruit, which consists of peltate scales fitting closely by their edges. In a subsequent part, the peculiarities, as regards the branch systems and leaves, of these cypresses (*Cupressus Lawsoniana, nootkatensis, thyoides, obtusa,* and *pisifera*) will be described, and may then be compared with those now given below for the four species of Thuya in cultivation.

In the discrimination of the Thuyas, in addition to the characters shown by the bark, mode of branching, and fruit, the primary and secondary axes of the branch-systems give good marks of distinction. These axes are markedly flattened in *Thuya occidentalis,* terete in the other species. In *Thuya orientalis* the branch-systems stand in vertical planes, the inner edges of which are directed towards the stem of the tree. In ordinary forms of the other three species they are arranged in horizontal planes. The leaves on the main axes in each species differ as follows :—

1. *Thuya plicata* :[1] widely spaced, long, ending in long, fine, free points, which are parallel to the axis; glands inconspicuous or absent. Under surface of the foliage usually marked with white streaks.

2. *Thuya japonica* : placed closely together, shoots ending in short, rigid, thick, triangular points, directed outwards at an acute angle; glands absent. Under surface of the foliage conspicuously marked with broad white streaks.

3. *Thuya occidentalis* : widely spaced, ending in long, fine points, which are parallel to the axis; glands raised, large and conspicuous on the flat leaves. Under surface of the foliage pale green; white streaks inconspicuous or absent.

4. *Thuya orientalis* : widely spaced, ending in short triangular free points, which are not rigid, and are directed slightly outwards at an acute angle : flat leaves marked by longitudinal glandular depressions. Under surface of the foliage pale green, without white streaks.

Thuya sutchuenensis, Franchet,[2] is a small tree occurring in north-east Szechuan in central China, where it was discovered by Père Farges growing at an altitude of 1400 feet. The branchlets are much flattened, thin in texture, and practically gland-less. Cones composed of 8 obovate scales, the apices of which are slightly thickened. This species has not been introduced into cultivation.

---

[1] This species exhales a peculiar aromatic odour, which is different from that of the other Thuyas.
[2] *Jour. de Bot.* 1899, p. 262. See also Masters in *Jour. Linn. Soc. (Bot.)* xxvi. 540.

THUYA PLICATA, Giant Thuya

Thuya plicata, D. Don in Lambert, *Pinus*, ed. 1, ii. 19 (1824) ; Masters, *Gard. Chron.* xxi. 214 ;
  figs. 69, 70, 71 (1897) ; Sudworth, *Check List Forest Trees U.S.* 31 (1898) ; Sargent, *Manual
  Trees N. America*, 75 (1905).
Thuya gigantea, Nuttall, *Jour. Philad. Acad.* vii. 52 (1834) ; Sargent, *Silva N. America*, x. 129, t.
  533 (1896) ; Kent, in Veitch's *Man. Coniferæ*, 239 (1900).
Thuya Menziesii, Douglas, ex Carrière, *Traité Gén. Conif.* 107 (1867).
Thuya Lobbi, Hort.
Thuya Craigiana, Hort. [*non* A. Murray, *Bot. Exped. Oregon*, 2 (1853)].

A lofty tree, attaining a height of 200 feet, with a trunk remarkably conical, the
base being broad and buttressed, sometimes girthing as much as 40 to 50 feet near
the ground.

Bark of the trunk fissuring longitudinally in narrow thick plates, which scale off,
leaving exposed the reddish brown cortex beneath.   On the branches, the bark only
begins to scale when they become old and thick.   Branches horizontal, ascending
towards their ends, forming in England a dense, narrow, pyramidal tree, usually
clothed to the base.

The 3-4 pinnate branch-systems, disposed in horizontal planes, have their main
axes terete and covered with long leaves ending in acute points which keep parallel to
the axes.   The glands on these leaves are inconspicuous or absent.   On the ultimate
axes the leaves are smaller, the flat ones scarcely glandular, and ending in mucronate
points ; the lateral ones keeled on the back, slightly curved, and ending in sharp
cartilaginous points.   On the lower surface of most branchlets the foliage is streaked
with white, some branchlets usually remaining uniformly green.

The male flowers are dark red in colour, cylindrical, and composed of about
6 decussate pairs of stamens.

The cones when ripe do not remain erect, but are deflected out of the plane
of the branchlets.   They are oblong, light brown in colour, and composed of 5 to 6
pairs of scales, of which the 2nd, 3rd, and 4th pairs are larger than the others, and
fertile.   The scales are oval or spathulate, with a rounded apex, from immediately
below which externally a small deltoid process is given off.   The seeds, 2 or 3 on
each fertile scale, are brown in colour, two-thirds the length of the scale, and
surrounded laterally by a scarious wing, which is deeply notched at its summit.

Seedling.[1]—The 2 cotyledons are linear, flat, acute at the apex, and slightly
tapering towards the base, supported on a terete caulicle, about $\frac{2}{5}$ inch long, which
ends in a long brown flexuose primary root giving off a few fibres.   The stem, terete
and smooth near the base, becomes ridged above by the decurrent leaf-bases.   The
first 4 true leaves are in opposite pairs, decussate with the cotyledons.   Above
these the stem gives off a number of whorls or pseudo-whorls of longer ($\frac{1}{2}$ inch)
sharply pointed leaves, dark green above and pale beneath, with markedly decurrent

---

[1] Figured in Lubbock, *Seedlings*, ii. 551, fig. 676 (1892), and Sargent, *loc. cit.* t. 533, fig. 12.

bases. After a few of these whorls lateral branches are given off, which sometimes bear a few acicular leaves at their bases. The lateral branches ramify and approach in character those of the adult plant, as the leaves are arranged decussately in 4 ranks. These leaves are variable, being acicular and loosely imbricated, or scale-like and closely imbricate. The branches are ascending, horizontal, or drooping, and are more or less flattened from above downwards.

HISTORY [1]

This tree was discovered by Née, who accompanied Malaspina in his voyage round the world during the years 1789 to 1794; and his specimen, gathered at Nootka Sound, is preserved in the Natural History Museum at South Kensington. It was referred to by James Donn, in *Hortus Cantab.* ed. 4 (1807), as *Thuya plicata*, without any description; and subsequently D. Don drew up from it the oldest description of the species under the same name. The *Thuya plicata* of gardens, which was early in cultivation, is a variety of *Thuya occidentalis*, and has no connection with the plant of Née.

Archibald Menzies, who accompanied Vancouver's expedition as botanist, gathered specimens also at Nootka Sound in 1795. Nuttall received specimens later from the Flathead river, on which he founded his description of the species as *Thuya gigantea*. It was introduced into cultivation [2] in 1853 by W. Lobb, and distributed from Veitch's nursery at Exeter as *Thuya Lobbi*, as at that time Nuttall's name *Thuya gigantea* was wrongly applied to *Libocedrus decurrens*, and Don's name, *Thuya plicata*, in a similar erroneous way, had gone into common use for a variety of *Thuya occidentalis*. Afterwards the tree became generally known in England as *Thuya gigantea*; and it is unfortunate that Don's name, *Thuya plicata*, must, following the law of priority, be substituted for a name so well known and so established as *gigantea*. This change of name has, however, been adopted in the *Kew Hand List of Conifers*, and by Sudworth and Sargent in North America, and on the whole it is now most convenient to adopt the name *Thuya plicata*.

(A. H.)

DISTRIBUTION

This tree is, next to the Douglas fir, the most important from an economic point of view in northern Oregon, Washington, and British Columbia.

It extends in the north as far as southern Alaska, in the east to the Cœur d'Alêne Mountains in Idaho and to north-western Montana, and in the south to Mendocino County in northern California. It is known as Cedar, or Red Cedar, and is found most abundantly on wet soils and in wet climates, ascending from sea level to an elevation, according to Sargent, of 6000 feet, where it becomes a low shrub.

---

[1] See Masters, in *Gard. Chron. loc. cit.*

[2] At the Royal Botanic Gardens, Edinburgh, there were in 1884 five trees of supposed *Thuya gigantea*, which were raised, it was said, from seed sent to Edinburgh by Jeffrey in 1851, while collecting for the Oregon Association. Three of these trees, according to Nicholson, were true *gigantea*, the other two being what is now known as *Thuya occidentalis*, var. *plicata*. See *Woods and Forests*, Feb. 27, and Mar. 19, 1884. These trees cannot now be identified.

It is scarce in the dry belt of country east of the Cascade Mountains, but common in the Selkirk and Gold ranges, though, so far as I know, it never extends to the eastern side of the Rocky Mountains.

On the coast and in Vancouver Island it attains an immense size. I have never measured trees more than 200 to 220 feet high, but Prof. Sheldon says that it attains 250 feet in Oregon, though no actual measurements are given.[1] As regards their girth, I have measured two trees which may have grown from the same root, so close do they stand together, one of which was 39, the other 25 feet at 5 feet from the ground. These stand on Mr. Barkley's farm in Vancouver Island, in swampy land near sea level, and are figured in Plate 56.[2] At over 2000 feet elevation in Oregon I measured another, also a twin tree, which was 30 feet in girth. Mr. Anderson states that he has seen Indian canoes 6 feet and more from the level of the gunwale to the bottom, hewn out of a log of this tree, such canoes being often 50 feet and more long. A hewn plank 5 feet wide by 15 feet long is in the museum at Victoria, B.C., and split boards, quite straight, 12 feet long and 15 inches wide, are made from it without difficulty.

The natural reproduction by seed was, wherever I saw it, very good, though in the densest shade the western hemlock seemed to have the advantage.

CULTIVATION

Wherever I have seen this tree growing in England and Scotland it is a vigorous, healthy tree of great beauty and promise, and one that I think is likely in fifty years or so to become a more valuable timber tree than the silver fir or spruce.

It has been stated in a report by Herr Bohm, in the March number of the *Zeitschrift für Forst. u. Jagdweser* for 1896,[3] that the parasitic fungus *Pestalozzia funerea* has done serious damage to the tree in North Germany, and statements to the same effect have been made elsewhere; but I can say that out of the thousands of this tree that I have raised from English seed and planted out in a bad soil and climate, I have never had any die from any disease whatever, and have found it an easier tree both to raise and to transplant than any other conifer. It will grow on almost any soil at the rate of at least one foot per annum, as in damp, cold bottoms where the spruce will hardly thrive, on the poor dry oolite soil of the Cotswold hills, and seems equally indifferent to wind, damp, and spring frosts.[4] It seldom loses its leader, is rarely blown down, endures heavy shade, and transplants both in

---

[1] In the Canadian Court of the Colonial Exhibition of 1886, there was shown a portion of a bole of this species, which was taken from a tree girthing 21 feet, and having a length of 250 feet. It came from British Columbia.—*Gard. Chron.* 1886, xxvi. 207.

[2] An illustration of a tree growing near Snoqualmie Falls on the Seattle and International Railway, Washington, was given in *The Pacific Rural Press* in 1897. This is said to have been 107 feet 7 inches round at the base, and was supposed to have been over 1000 years old, but we know of no good evidence that it ever attains so great an age as this.

[3] *Cf.* A. C. Forbes, *Gard. Chron.* 1896, xix. 554.

[4] The very severe frost on May 20-21, 1905, when 10°-15° of frost were registered in many places, which killed many young beech trees in low situations at Colesborne, and checked the young growth considerably, killed none except a few of the weakest Thuyas which were freshly transplanted; but the autumn frost of the following October, when the trees were still in growth, seems to have done more harm, though the young trees did not die till the following spring.

early autumn and late spring with great readiness. It has, therefore, every good quality a forest tree can have, except the as yet unproved one of cleaning its trunk from branches without pruning.

And as this has not yet been properly tested by thick planting, I venture to say that there is no conifer better worthy of an extensive trial as a timber tree for such purposes as the larch is now used, and especially for fencing posts, for which its remarkable durability in the ground seems to make it most valuable.[1]

I should therefore recommend that this tree should be planted at distances of 6 to 8 feet apart in situations where larch will not thrive, and not thinned as long as the trees keep healthy.

In the New England states it is not hardy enough to live in many places, but Professor Sargent tells me that a variety raised from seed from the Coeur d'Alène mountains in northern Idaho is hardy at Boston, where the form from the Pacific coast is tender, just as in the case of the Douglas fir.

No reliable tests, so far as I know, have yet been made in England or America as to the breaking strain and strength of this wood, but Sheldon states that it is used for telegraph posts in Oregon, and though its branches die off so slowly that the home-grown timber may probably be knotty, it is certainly not worse in this respect than spruce, to which I should consider it in every respect a superior forest tree.

The seed usually ripens about the end of October, and is very freely produced in most seasons. It soon sheds when ripe, and should be sown in boxes or in the open ground in early spring. I have tried both plans with great success, and find it best to plant the seedlings at two years old in nursery lines, and plant out the trees finally either in the early autumn or spring, when the deaths will be very small if the roots are not allowed to dry before planting.

There is very little variation among the seedlings, which grow rapidly in moist soil, and are less liable to suffer from spring frost than most trees, though if planted in mid-winter the tops are liable to die back.

There is no reason why this tree should not be sold in nurseries at the price of spruce except the absence of a regular demand, as it can be got up to a proper size for planting in two years less time.

The tree seeds itself very rapidly on sandy soil in many parts of the west and south of England, though liable to be thrown out of the ground by frost during the first year, and often destroyed by rabbits. On the lower greensand at Blackmoor, Hants, self-sown seedlings were quite numerous, both of this tree and of many other conifers, but rabbits are not allowed here, and both Lord and Lady Selborne take great interest in self-sown seedlings.

REMARKABLE TREES

The giant Thuya has not been long enough in cultivation to show whether it

---

[1] I have recently been shown by Mr. Molyneux a plantation of *Thuya gigantea* and larch called Mays hill, made by him in 1888 on poor, heavy wheat land overlying chalk at Swanmore Park, Hants, the seat of W. H. Myers, Esq., M.P. Here the Thuyas have completely outgrown the larch, and in many cases suppressed them, and are 15 to 20 feet high, and quite healthy ; whereas where the larch were planted alone in the same place they are diseased and sickly.

will attain the same dimensions that it does in America, but there are many trees which are already 60 to 70 feet in height at less than fifty years from seed.

By far the finest that I have seen or heard of are at Fonthill Abbey, Wilts, the residence of Lady Octavia Shaw-Stewart, which were raised in the late Duke of Westminster's gardens at Eaton Hall from seeds collected for Lord Stalbridge in 1860. Here, on a bed of greensand at an elevation of 400 to 500 feet, well sheltered from wind, are growing some of the finest and best grown conifers in Great Britain. In a group of three Thuyas, the middle one measured in 1906, as nearly as I could ascertain, not less than 90 and probably 95 feet in height by 10 feet in girth, and already began to show the buttressed trunk which is so characteristic in its native country. The other two trees were not much less in size, and all were a picture of health and symmetry (Plate 57).

The next tallest that we know of is a tree at Albury Park, the Surrey seat of the Duke of Northumberland. This was measured by Henry in 1904, and by myself in 1905, but owing to the way in which it is shut in by other trees it is difficult to measure accurately, and though the late Mr. Leach, the head gardener at Albury, and Dr. Henry both considered it about 90 feet high, I should not like to say that it is over 80, with a girth of 7 feet 6 inches. It is, however, a very healthy and vigorous tree, and growing fast, and the Duke's agent and gardener both hold a very high opinion of the probable value of the tree for timber, and are planting it largely on the estate. See *Gard. Chron.* Jan. 30, 1892, where an account is given of the trees at Albury in which Mr. Leach is quoted as saying : " If I had 1000 acres to plant with trees that would give the most remunerative return in a given time, the above would be my mainstay."

Sir Charles Strickland, one of the oldest and most experienced planters in England, also has a high opinion of this tree, and is quoted as follows by Mr. A. D. Webster in an article on this tree in *Trans. Scottish Arb. Soc.* vol. xii. p. 343 :— " There is a hillside here (Hildenley, Yorkshire), with a thin soil upon limestone rock, which I planted two or three times over with very small success—chiefly, I believe, on account of the extreme dryness of the site. The Thuya grows there with great vigour, and I have scarcely lost one of those planted. Among the other merits of this Thuya is the ease with which it may be transplanted, owing to its having bushy, fibrous roots, instead of the long tangles which larch and many other conifers have." I saw this plantation in 1905, and though the situation is too dry for Thuya to grow to any size, it bears out Sir Charles's good opinion. He has continued to raise the tree largely from his own seed, and is planting them largely at 5 feet apart, without mixture.

At Castlehill, North Devon, the seat of Earl Fortescue, there are also very fine specimens of *Thuya plicata*. The best is growing in a quarry in a well-sheltered place, but on dry, rocky ground. It measured in April 1905 about 74 feet high by 5 feet 11 inches in girth, and bids fair to become a noble tree.

At Fulmodestone, Norfolk, there are two trees, planted in 1863, which measured in 1905, 67 feet by 7 feet, and 61 feet by 6 feet 8 inches, and have natural seedlings around them.

At Coolhurst, near Horsham, Mr. C. Scrace Dickens showed me a very fine and symmetrical tree 75½ feet high by 5½ in girth, and only 8 yards in the spread of its branches.

At many places in the south-west of England trees of from 65 to 70 feet are growing of which the following are the best we have measured ourselves :—Linton Park, Kent, 70 feet by 7 feet 1 inch in 1902 ; Dropmore, Bucks, 68 feet by 6 feet 10 inches in 1905 ; Killerton, Devonshire, 68 feet by 7 feet 10 inches in 1905 ; Bicton, Devonshire, 70 feet by 8 feet 2 inches in 1902 ; Blackmoor, Hants, 60 feet by 6 feet.

In Wales a tree at Hafodunos measured 65 feet 6 inches by 9 feet 7 inches in 1904, with natural seedlings a few feet from its base on the stump of an old tree ; at Welfield, near Builth, the seat of E. D. Thomas, Esq., a tree 68 feet high and 6½ feet in girth was flourishing on the Llandilo slate formation ; and at Penrhyn Castle Mr. Richards showed me a well-shaped and healthy young tree about 50 feet high, one of fifty which had been transplanted when about 18 feet high, only one of which died after being moved.

In Scotland *Thuya plicata* flourishes in the south and west, as well as in England. At Inverary Castle a tree only 25 feet high in 1892 is now over 60. At Poltalloch there are many, of which one in 1905 was 65 feet by 7 feet 2 inches. As far north as Gordon Castle it grows well, and at most of the places from which reports were sent to the Conifer conference in 1892 it is spoken of as healthy and vigorous. At Murthly, Scone, and Castle Menzies, I have seen fine trees, but have not measured any of remarkable size.

At Monreith, Dumfriesshire, the seat of Sir Herbert Maxwell, Bart., who has a high opinion of this tree, a large number have been raised from seed and planted out, but are as yet too young to measure.

At Benmore, near Dunoon in Argyllshire, the property of H. J. Younger, Esq., where there are very interesting plantations of several kinds of exotic conifers made in the winter of 1878-79, Thuya, when mixed with the common larch and Douglas fir on a steep hillside at 250 to 500 feet above sea-level, is now being suppressed by these species, which grow more vigorously. However, in one part of the plantation, near Ardbeg, at only 50 feet above sea-level and in fairly good soil, the Thuya was holding its own fairly well with the Douglas, and had attained, at twenty-four years old, 50 feet in height with clean stems varying from 25 to 38 inches in girth at 5 feet from the ground. Near Kilmun, on the same property, there is now, according to the forester, about 1½ acres of Thuya, which has been planted mixed with larch. The larch has been cut out, and the whole area is now pure Thuya, with clean stems larger in size than in the other parts of the plantations where it occurs mixed with Douglas fir.[1]

In Ireland the best trees we know of are at Castlewellan, co. Down, 65 feet in 1903 ; Hamwood, co. Meath, 71 feet by 6 feet 3 inches in 1904 ; Churchill, co. Armagh, 68 feet by 5 feet 10 inches in 1904 ; Adare, co. Limerick, 71 feet by 7 feet 7 inches in 1903.

---

[1] We are indebted to Mr. Angus Cameron, factor for the property, and to Mr. J. M. Stewart, forester, for further particulars of these plantations, for which we cannot now find space.

At Dartrey, in co. Monaghan, the Earl of Dartrey planted in 1882 a considerable area of slightly hilly ground with a mixture of larch, spruce, Douglas fir, and Thuya. In 1904, twenty-two years after planting, of the four species, all grown densely under the same conditions, the Thuya had made the most timber, the trees averaging 40 to 50 feet in height by 4½ feet in girth. The Douglas fir was slightly taller, but not so stout in the stem, averaging about 3½ feet in girth. The Earl of Dartrey speaks very highly of the timber of Thuya, which he considers to be superior to that of the best larch.

At Brockley Park, Queen's Co., the residence of Mr. Wm. Young, there are trees growing on light soil on limestone, which have made 40 cubic feet of timber in 30 years, and 50 feet in 35 years. The tallest tree, 30 years old, was in 1906 64 feet high by 7 feet 9 inches at a foot from the ground, and 3½ feet girth at 24 feet up; and its branches were 105 feet in circumference.

TIMBER

Sargent says, *Garden and Forest*, iv. p. 109: "The wood is very valuable; it is light, soft, and easily worked, and so durable in contact with the ground, or when exposed to the elements, that no one has ever known it long enough to see it decay."

The great value of the cedar for shingle-making has long been known, and several instances were mentioned by reliable people in Vancouver Island of hand-made shingles, or "shakes" as they are called, remaining good 40 to 50 years on roofs without decaying in the wet climate of this island.

They are now manufactured on a very large scale by machinery in all the Puget Sound mills, and exported largely to the middle and eastern states in neat bundles, and I have no doubt that, if carefully selected and laid, such shingles would be very suitable for roofing in England. Sargent says, *Garden and Forest*, iv. p. 242, "that nearly 100 mills were in 1891 exclusively devoted to making Red Cedar shingles, and that the combined output of half of these operated by one company was 3,500,000 per diem. They are now supplanting the Pine shingle of Michigan, the Cypress shingle of the south, and the Redwood shingle of California."

As a rule in the American forests, they begin to decay at the heart long before they attain their full growth, and the trunk seems to continue growing round the hollow centre for an almost indefinite time, as in the case of the yew. On drier land it keeps sound longer, and if cut when 2 to 3 feet in diameter the wood is probably at its best. It resists decay for an immense time when fallen.

For inside finish the wood is excellent, though not hard enough for flooring and wainscot, or strong enough for joists. For ceiling and panelling it is most ornamental when well cut, as I saw in the Hotel at Duncan's, Vancouver Island.

Mr. Stewart has found at Benmore that it is very suitable for all estate purposes, and prefers it to larch for planking and fencing, as he finds it less liable to warp and crack.   (H. J. E.)

THUYA OCCIDENTALIS, Western Arbor Vitæ

Thuya occidentalis, Linnæus, *Sp. Pl.* 1002 (1753); Loudon, *Arb. et Frut. Brit.* iv. 2454 (1838);
  Sargent, *Silva N. America*, x. 126, t. 532 (1896), and *Manual Trees N. America*, 74 (1905);
  Masters, *Gard. Chron.* xxi. 213, figs. 67, 68, and 258, fig. 86 (1897); Kent, in Veitch's *Man.
  Conif.* 244 (1900).
Thuya plicata, Hort. (*non* Don).

A tree, attaining a height of 50 to 60 feet, with a stout and buttressed trunk, some-times 6 feet in diameter. It often divides near the base into two or three stems. In England the branches, short and spreading, form a tree pyramidal in outline, which is not so dense in foliage as *Thuya plicata*. Bark of the trunk scaling off in thin papery rolls, but not so freely or so finely as in *Thuya japonica*. The branches when of no great size begin to show scaly bark.

The branch systems are disposed in horizontal planes, resembling those of *Thuya plicata*; but their main axes are flattened, being compressed from below upwards, while the leaves are shorter than in that species, ending in similar long points. The flat leaves on the main axes are studded with conspicuous large circular elevated glands. The smaller leaves on the ultimate branchlets vary as regards the presence or absence of glands; the lateral pairs are shorter than and not so acutely pointed as in *Thuya plicata*. The foliage is dark green above, pale green and not marked with white streaks below.

The male flowers, minute and globose, are composed of three decussate pairs of stamens. The female flowers are yellow.

The cones become deflected when ripe, as in *Thuya plicata*. They are oblong, light brown, and composed of 4 to 5 pairs of scales, of which the 2nd and 3rd pairs are larger than the others, and fertile. The scales are ovate or spathulate, ending in a rounded or acute apex, with a minute external process, which is generally much less developed than is the case in *Thuya plicata*. The seeds, usually two on each fertile scale, are scarcely distinguishable from those of the last-named species.

Seedling.[1]—Cotyledons as in *Thuya plicata*. The caulicle and stem are quad-rangular. The first two true leaves are opposite, spreading, and similar to the cotyledons, though smaller. These are followed by 5 or more whorls or pseudo-whorls, each of three similar leaves, linear, acute, and sessile. The ultimate leaves are opposite, decussate, and adnate for the greater part of their length to branchlets, which are flattened from above downwards.

Varieties

Few trees, except *Cupressus Lawsoniana*, show a greater tendency to varia-tion in the seed-bed. Sargent says that if anyone will sow a quantity of seed he will be sure to find forms among the seedlings as novel and as interesting

---

[1] See Lubbock, *Seedlings*, ii. 548, 560 (1892).

as any now in cultivation. Many of the varieties only show their distinctive characteristics when young, and soon grow up into the normal form. Beissner gives as many as forty varieties; but it is doubtful if all these are recognisable. Those commonly met with in cultivation in this country are enumerated below :—

    1. Var. *ericoides*.[1]

        Retinospora dubia, Carrière, *Conif.* ed. 2, p. 141.

A form in which the seedling foliage is fixed and preserved. It is a dwarf, compact, rounded, or somewhat pyramidal shrub, with slender branchlets, on which the leaves, heath-like in appearance, are borne in distant decussate pairs. They are spreading, linear, and soft in texture, becoming brown in winter. This shrub resembles *Cupressus pisifera*, var. *squarrosa*; but in the latter the leaves are much whiter on both surfaces, and do not brown in winter. The latter also attains a much larger size, and often becomes a large shrub or small tree.

    2. Var. *Ellwangeriana*.

        Retinospora Ellwangeriana, Carrière, *Rev. Hort.* 1869, p. 349.

This is a transition form, in which both kinds of foliage, seedling and adult, appear on the shrub, which may attain a considerable size. There is no regularity in the distribution of the two kinds of leaves; but in shrubs at Kew of this variety the juvenile foliage persists on branchlets in the interior shaded parts, the external branchlets having adult foliage.

It was probably this form which M'Nab[2] mentions as having seen in 1866 in quantity in the nursery of Messrs. P. Lawson and Sons, who had received it from Messrs. Ellwanger and Barry of America under the name of Tom-Thumb Arbor Vitæ. M'Nab states that the heath-like leaves have a slight smell of juniper, while the other foliage has the odour of ordinary *Thuya occidentalis*.

    3. Var. *plicata*, Masters, *Gard. Chron.* xxi. 258, fig. 86 (1897).

        Thuya plicata, Parlatore, *D.C. Prod.* xvi. 457.

A tree differing from the type in the branch-systems tending to assume the vertical plane, being curved so that the ultimate branchlets lie in different planes. The foliage is conspicuously glandular, the lateral leaves being flattened, so that they become almost like the median ones in appearance. According to Kent the foliage shows a brownish tint.

This variety was long considered to be a distinct species; but it is only a seedling of *Thuya occidentalis*, with which it agrees in cones and in general character of the leaves.

    4. Var. *Wareana*. This only differs from the last in the colour of the foliage, which is a deep green without any brown tinge. It was raised by Mr. Ware of Coventry.[3] According to Masters[4] it has larger leaves than var. *plicata*, and corresponds very closely with native specimens of *Thuya occidentalis* gathered at Niagara.

---

[1] A plant of this variety growing into the mature form at Meehan's nursery, Germantown, U.S., showed that it was only a juvenile state of *Thuya occidentalis*.—*Garden and Forest*, 1893, p. 378.

[2] *Trans. Edin. Bot. Soc.* ix. 61, fig. (1868).

[3] Gordon, *Pinetum*, ed. 2, p. 409.

[4] *Gard. Chron.* xxi. 258 (1897).

5. Var. *dumosa*. A dwarf shrub, with the foliage and branchlets of var. *plicata*.

6. Var. *pendula*. A shrub with pendulous branches and branchlets.

7. Var. *erecta*. Branches slender and erect. In var. *erecta viridis* the foliage is dark green and shining on the upper surface. It originated in Messrs. Paul's nurseries at Cheshunt.[1]

8. Var. *Späthi*. A monstrous form, with seedling foliage on the younger branchlets, older branchlets being tetragonal, and clothed with sharp-pointed adult leaves.

9. Various forms occur with coloured foliage, as *lutea, aurea, vervæneana*, etc.

Thuya occidentalis was probably the first American tree cultivated in Europe. Belon[2] describes it as occurring in a garden at Paris about the middle of the sixteenth century. It was introduced into England prior to 1597, as it is mentioned by Gerard in his *Herball* published in that year.   (A. H.)

DISTRIBUTION, ETC.

According to Sargent, *Thuya occidentalis* frequently forms nearly impenetrable forests on swampy ground, or occupies the rocky banks of streams from Nova Scotia and New Brunswick, north-westward to Cedar Lake at the mouth of the Saskatchewan, and southward through the northern states to southern New Hampshire, central Massachusetts and New York, northern Pennsylvania, central Michigan, northern Illinois, and central Minnesota, and along the high Alleghany mountains to southern Virginia and north-eastern Tennessee; very common in the north, less abundant and of smaller size southward; on the southern Alleghany mountains only at high elevations.

Mr. James M. Macoun says of this tree in his excellent pamphlet, *The Forest Wealth of Canada* (Ottawa, 1904), that the white cedar, as it is there usually called —though in New England this name is always given to *Cupressus thyoides*—is very rare in Nova Scotia, but abundant throughout New Brunswick and Ontario. It grows to a considerable height, but seldom exceeds 2 feet in diameter. The wood is soft and not strong, and has never been much used for timber, but is unexcelled for shingles. It is chiefly used for fence rails and posts, railway ties, and telegraph posts. No other wood is used in any quantity for telegraph poles in Ontario and Quebec. It is very durable in contact with the soil or when exposed to the weather.

I saw the tree abundantly in wet swamps and also on dry ground near Ottawa, where, in Rockcliff Park, good though not large trees of it may be seen, the best having all been cut out for telegraph poles. On dry, rocky ground the tree grows freely from the stool, and in wet places in the woods reproduces abundantly from seed, which was ripe at the end of September, and, as usual in the forests of Canada, germinates and grows best when it falls on a rotten log.

---

[1] *Gard. Chron.* xiv. 213 (1880).   [2] Belon, *De Arboribus Coniferis*, p. 13 (1553).

REMARKABLE TREES

Thuya occidentalis never attains to a considerable size when planted in this country. There is a specimen at White Knights, near Reading, of great age, which is now dying at the top. According to the gardener there it has not made any growth for the last thirty-five years. It measured in 1904, 41 feet in height by 4 feet in girth. At Stratton Strawless, Norfolk, there is also a specimen of considerable age, remarkable for the pendulous habit of the branches, which is 35 feet in height. There are more large specimens at Belton Park than at any other place I know in England, the largest I have measured being 41 feet by 3 feet 9 inches. Henry, however, in 1904 measured one at Arley Castle as tall, which divides into three stems near the ground, where it measures 7 feet 6 inches in girth. At Auchendrane, Ayrshire, Renwick measured a tree in 1902—which, according to a specimen procured by him in 1906, was *Thuya occidentalis*—as 42 feet high by 6 feet 8 inches in girth, with a bole of 12 feet.

It seems to be one of the best conifers for making shelter hedges in gardens, as it stands clipping well, and for this purpose may be relied on to attain 15 to 20 feet in height in any fair soil. As it grows slowly at first when raised from seed, it is usually propagated by cuttings.　　　　　　　　　　　　　　　　　　(H. J. E.)

THUYA JAPONICA, Japanese Thuya

Thuya japonica, Maximowicz, *Mél. Biol.* i. 26 (1866); Masters, *Jour. Linn. Soc.* (*Bot.*) xviii. 486 (1881), and *Gard. Chron.* xxi. 258, fig. 87 (1897); *Revue Horticole*, 1896, p. 160; Kent, in Veitch's *Man. Coniferæ*, 244 (1900); Shirasawa, *Icon. des Essences forestières du Japon*, 28, t. xi. 18-34 (1900).
Thuya Standishii, Carrière, *Traité Gén. Conif.* 108 (1867).
Thuya gigantea, var. *japonica*, Franchet et Savatier, *Enum. Pl. Jap.* i. 469 (1875).
Thujopsis Standishii, Gordon, *Pin. Suppl.* 100 (1862).

A tree attaining, according to Shirasawa, a height of 90 feet in Japan, with a tapering stem, open in habit as cultivated in England, and not forming such a dense pyramid as *Thuya plicata*. Bark of the trunk scaling off in very narrow longitudinal papery strips. The bark commences to scale on young branches of less than a half inch in diameter. The branches curve upwards towards their extremities.

The branch-systems, 3-4 pinnate, are disposed in horizontal planes, which droop at their outer extremities. Primary axes terete, with leaves densely crowded, all the four sets ending in short, rigid, thick, free points, glands being absent. The leaves on the ultimate branchlets are obtuse, and not acutely pointed as in *Thuya plicata*; and glands may be present or absent on the flat leaves. The foliage is light green above, while on the under surface there are whitish streaks, somewhat triangular in outline, which exceed in area the greener parts.

Male flowers cylindrical, with 6 decussate pairs of stamens. The cones are deflected, ovoid, and composed of 5 to 6 pairs of scales, of which the second and third pairs are larger than the others and fertile. The scales are broadly oval, with a rounded apex, from below which externally is given off a short, broad, triangular process, projecting from the scale at right angles or nearly so. The seeds, three to each fertile scale, and nearly equal to it in length, differ considerably from those of *Thuya plicata* and *Thuya occidentalis*, the wing being narrow, not so scarious in texture, entire, and not notched at the summit.

Fortune discovered *Thuya japonica* in cultivation around Tokyo in 1860, and sent home seeds of it to the nursery of Mr. Standish at Ascot, who distributed plants under the name of *Thujopsis Standishii*. Maximowicz, who had also seen it cultivated at Tokyo, gave the species its first authoritative name in 1861. Maries found it growing wild on the mountains of Nikko, in central Japan, in 1877. Sargent,[1] who, in company with James H. Veitch, met with a few solitary specimens on the shores of Lake Yumoto in these mountains, at 4000 feet altitude, describes it as a small pyramidal tree of 20 to 30 feet high, of open and graceful habit, with pale green foliage and bright red bark. Shirasawa, however, states that it attains a height of 90 feet, with a diameter of stem of nearly 6 feet; and that it grows in the central chain of Hondo, in the mountains of Kaga, Hida, and Shinano, at elevations of 2000 to 6600 feet. The stem, according to Shirasawa, is often twisted, and gives off great wide-spreading branches. (A. H.)

[1] *Garden and Forest*, 1893, p. 442, and 1897, p. 441.

According to Komaror, *Floræ Manshuriæ*, i. 206 (1901), *Thuya japonica* grows wild abundantly in northern Corea in the Samsu district, but was not observed by him in Manchuria or elsewhere on the mainland.

This tree is not, so far as I saw, as common in Japan, where it is called Nezuko, as *Cupressus obtusa* or *C. pisifera*, though it is said by Goto[1] to be found in the provinces of Yamato, Bungo, Satsuma, Omi, Iwashiro, Shimotsuke, and Uzen, at an elevation of from about 3000 to 6000 feet.

The only place where I saw it wild was at Yumoto, above Nikko, where it was scattered in mixed forest with Tsuga, Thujopsis, birches, and other deciduous trees, and it is said to be never found in unmixed woods.   At Koyasan I found small trees of it, perhaps planted, and brought away a seedling, which is now living at Colesborne.

At Atera, in the Kisogawa district, the forester told me that it grows best as a young tree in shade, and that where *Cupressus obtusa* has been felled it often comes up from seed.   It does not attain very large dimensions, so far as I could learn, and is not considered a tree of much economic importance.

The timber is light and used for carpentry.   It sometimes has a very pretty figure, and in old trees is of a pale grey colour, though perhaps this is only assumed by trees which were dead before cutting.   It is cut into thin boards, and used for ceilings and other inside work, and is said to cost about 2d. per square foot in the board at Tokyo, and to make very durable shingles.

In Great Britain the tree seems to grow slowly, and is not common in gardens. The largest I have seen is a grafted and very spreading tree in Mr. W. H. Griffiths' garden at Campden, Gloucestershire, which is about 25 feet by $2\frac{1}{2}$ feet, and probably one of the oldest in England.   It has produced fertile seeds from which plants have been raised.   The largest recorded at the Conifer Conference was at Dalkeith Palace, where it was 15 feet high in 1891.   A tree at Kilmacurragh, co. Wicklow, Ireland, was 24 feet by 2 feet 4 inches in 1906, and bears fruit.   Another at Castlewellan measured 25 feet high in the same year.   (H. J. E.)

---

[1] *Forestry of Japan* (1904).

THUYA ORIENTALIS, Chinese Arbor Vitæ

Thuya orientalis, Linnæus, *Sp. Pl.* 1002 (1753); Loudon, *Arb. et Frut. Brit.* iv. 2459 (1838);
Masters, *Jour. Linn. Soc.* (*Bot.*) xviii. 488; Kent, in Veitch's *Man. Coniferæ*, 248 (1900).
Biota orientalis, Endlicher, *Syn. Conif.* 47 (1847).

A tree or dense shrub, with the trunk often branching into several stems from
near the base. Bark of trunk thin, reddish brown, and separating in longitudinal
papery scales. The bark begins to scale on branches which are about a half inch in
thickness. The branches are ascending, becoming tortuose at their extremities,
and giving off more or less equal-sided branch-systems, which are disposed in
vertical planes, with their inner edge directed towards the stem of the tree. These
are finer and more closely ramified than in the preceding species. Their main axes
are terete; bearing median leaves, marked by a glandular longitudinal depression,
and ending in triangular free points (not appressed to the axis); and lateral leaves,
ending in similar but longer free points, which are thickened at the part where they
become free and reflected away from the axis. The leaves on the ultimate branchlets
are closely imbricated, appressed to the stem, and marked with longitudinal depressions.

The male flowers are globose and composed of 4 decussate pairs of stamens.

The cones[1] are erect and ovoid, fleshy and bluish before ripening, but
ultimately becoming dry and woody, the scales gaping widely. Scales, usually
3 pairs (occasionally a fourth pair, sterile and much reduced, appears at the base),
the two lowest fertile, the uppermost pair aborted and sterile: ovate, obtuse, thick,
and ligneous, bearing externally below the apex a hooked process. The seeds,
2 on each scale, are large, ovoid, without wings, brown in colour, with a white, large,
oblong hilum.

The seedling[2] resembles that of the other species of Thuya, except that the
cotyledons are much larger, about an inch in length.

Varieties

A great number of varieties of this species have been obtained. The most
remarkable of these are :—

1. Var. *pendula*, Masters, *Jour. R. Hort. Soc.* xiv. 252.

> *Thuya pendula*, Lambert, *Genus Pinus*, ed. 2, ii. 115 t. 52; Siebold et Zuccarini, *Fl. Jap.*
> ii. 30 t. 117.
> *Thuya filiformis*, Lindley, *Bot. Reg.* xxviii. t. 20 (1842).
> *Biota pendula*, Endlicher, *Syn. Conif.* 49.
> *Cupressus pendula*, Thunberg, *Fl. Jap.* 265.

[1] The cones ripen in one year, but frequently in England retain their seed till the spring of the following year.
[2] Tubeuf, *Samen, Früchte, u. Keimlinge*, 104, fig. 144 (1891).

A shrub, with a straight trunk, bare of branches below. The branchlets, numerous, long, flexile, cord - like, unbranched or only slightly branched, are produced in irregular fascicles of 5 to 20 or more at irregular intervals along the branches. They are slender and pendent, and bear leaves distantly placed in 4 rows in decussate pairs. The leaves, broadly decurrent at the base and long acuminate at the apex, spread out from the branchlets at an acute angle. Cones are occasionally borne, which are like those of the type.[1]   There is a specimen at Kew of a plant raised from seed of this variety, which is ordinary *Thuya orientalis*. It was sent from the Botanic Garden at Turin by Mr. Hanbury in 1860.

There are several forms of this variety, differing in habit and length of leaves ; in one the branchlets are tetragonal.

This shrub was first observed by Thunberg in Japan, and specimens were collected near Yokohama by Maximowicz. It was also met with by Fortune in China, and has been raised in Europe.

2. Var. *decussata*.
Retinospora juniperoides, Carrière, *Conif.* ed. 2, p. 140.

A low shrub, with erect stems and branches, bearing foliage like that of the seedling. The leaves are in 4 rows in decussate pairs, spreading, and resembling those of a juniper, except that the points are not prickly. They are greyish green in summer, changing to brown in winter.

3. Var. *Meldensis*.
Biota Meldensis, Lawson, in Gordon, *Pinetum*, 37.

A small tree with ascending flexible branches. It is a transition form, bearing acute acicular spreading leaves like that of the seedling, and occasionally leaves of the adult character. The leaves are bluish green, changing to brown in the winter. This plant was raised from seeds of *Thuya orientalis* gathered in the cemetery of Trilbardoux near Meaux in France ; and for a long time was supposed to be a cross between *Thuya orientalis* and *Juniperus virginiana*.

4. Var. *intermedia*.
Biota orientalis intermedia, Carrière, *Man. des Pl.* iv. 322.

This is also a transition form. It is a shrub with elongated pendent branchlets, the ramifications of which arise from all sides of the axis, not remaining in one plane. There are two kinds of leaves, those towards the ends of the branchlets resembling the adult foliage of *Thuya orientalis*, while those on older parts are spreading, arranged in decussate pairs, oval-lanceolate, decurrent at the base, and acute at the apex. In Var. *funiculata*, if it is in reality distinguishable, there appears to be a larger proportion of adult foliage.

Many other varieties have been described : some of peculiar habit, as *gracilis* and *pyramidalis*, which are fastigiate ; others with coloured or variegated foliage, as *aurea*, *argenteo - variegata*, *aureo - variegata*. Var. *ericoides* of this species closely

---

[1] At Barton, a shrub of this variety produced cones, which had very long hooked processes on the scales (Bunbury, *Arboretum Notes*, 153).

resembles the variety of the same name belonging to *Thuya occidentalis*; the latter is slightly whiter on both surfaces of the leaves.

DISTRIBUTION, ETC.

Thuya orientalis occurs wild in the mountains of north China. It is common in the hills west of Pekin, where Fortune[1] observed trees of a large size, 50 or 60 feet in height. Elsewhere in China it is only met with planted in cemeteries and temple grounds. It has been known to the Chinese from the earliest times as the *Poh* or *Peh* tree, and is mentioned in their classical books; it was planted around the graves of feudal princes, and its wood was used for making the coffins of great officials. The tree was introduced into Japan from China at an early period, probably like so many other Chinese plants, by the Buddhist missionaries. Japanese botanists are all agreed that it is not indigenous in Japan. Various other regions have been mentioned as being the home of *Thuya orientalis*, as Siberia, Turkestan, Himalayas, etc.; but specimens collected in these countries are undoubtedly from cultivated trees. The tree is mentioned by Gmelin in his *Flora Siberica*, i. 182 (1747); but only as occurring between Kiachta and Peking. Ledebour[2] denies its existence in any part of Siberia.

Thuya orientalis was first grown in Europe at Leyden, some time before 1737, when Linnæus[3] described the plant as *Thuya strobilis uncinatis squamis reflexa acuminatis*. Royen, who sent a specimen to Linnæus, mentions the species with considerable details in his account[4] of the plants that were cultivated at that time in the Botanic Garden at Leyden; but his promised account of the history of its introduction apparently never was published. It is possible that it was raised from seed sent home by the Dutch from Japan, as Kaempfer, who travelled in that country from 1690 to 1692, collected specimens of *Thuya orientalis* which are still preserved in the Natural History Museum at South Kensington.[5] Seeds were also soon afterwards sent to Paris by the missionaries in north China.[6] The earliest account of it in England occurs in a letter dated February 1, 1743, from the Duke of Richmond to Collinson, as follows :—" I am sorry to find by Miller that I am not likely to have the Chinese Thuya. I own, if it belonged to anybody that would sell it, I should be foolish enough to offer ten guineas for it, because it is the only one in England that can match that which I have already." It was cultivated early by Miller[7] in the Physic Garden at Chelsea.

Thuya orientalis never attains in this country any considerable dimensions. It ripens good seed; and at Kew, on a wall near the Director's office, may be seen a

---

[1] *Yedo and Peking*, 307, 382 (1863). Fortune supposed that the wild tree in north China was distinct from that cultivated near Shanghai; but there is no doubt that the trees, which attain a great size in the hills west of Peking, are ordinary *Thuya orientalis*.

[2] *Comment. in Gmelini Fl. Sibericam*, 60 (1841). [3] *Hort. Cliff*. 449 (1737).

[4] *Floræ Leydensis Prodromus*, 87 (1740).

[5] I have seen these specimens. See Salisbury, *Coniferous Plants of Kaempfer*, in *Jour. Science and Arts*, ii. 313 (1817). Kaempfer does not mention the plant in his *Amœnitates Exoticæ*.

[6] See Miller, *Gard. Dict.* ed. 6 (1752), and ed. 8 (1768), *sub* " Thuya."

[7] *Cf.* Aiton, *Hort. Kew*. iii. 371 (1789).

young tree which originated from a seed probably carried there by a bird from a tree in the gardens.

In the garden at Hampton Court, Herefordshire, there are a pair of fine specimens about 40 feet high, and about 7 feet round at the base, where they divide into several stems which have been formed into an arch over the path, and in most old gardens trees of 25 to 35 feet may be found, but, like *T. japonica* and *T. occidentalis*, it must be looked on as an ornamental shrub rather than a timber tree.

(A. H.)

Printed by R. & R. CLARK, LIMITED, *Edinburgh.*

BEECH DRIVE AT CIRENCESTER

PLATE 2.

BEECH AVENUE AT WATFORD

PLATE 3.

QUEEN BEECH AT ASHRIDGE

PLATE 4.

INARCHED BEECH AT ASHRIDGE

PLATE 5.

BEECH, WITH BURR, AT ASHRIDGE

PLATE 6.

PLATE 7.

GIANT BEECH AT CORNBURY

PLATE 8.

BEECH AT NEWBATTLE

PLATE 9.

TRUNK OF NEWBATTLE BEECH

PLATE 10.

BEECH AT GORDON CASTLE

PLATE 11.

BEECH HEDGE AT MEIKLEOUR

PLATE 12.

KING BEECH AT KNOLE

PLATE 13.

AILANTHUS AT BROOM HOUSE

PLATE 14.

AILANTHUS AT BELTON

B

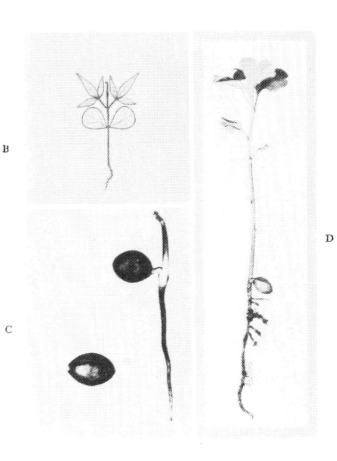

C

D

F

PLATE 15.

SEEDLINGS

A B, AILANTHUS.   C D, GINKGO.   E F, ARAUCARIA.

PLATE 16.

SOPHORA AT CAMBRIDGE

ARAUCARIAS IN CHILE

PLATE 17.

PLATE 18.

ARAUCARIA FOREST IN CHILE

PLATE 19.

ARAUCARIA AT DROPMORE

PLATE 20.

ARAUCARIAS AT BEAUPORT

PLATE 21.

MAIDENHAIR TREE AT KEW

PLATE 22.

MAIDENHAIR TREE AT FROGMORE

PLATE 23.

MAIDENHAIR TREE IN CHINA

PLATE 24.

TULIP TREE IN NORTH CAROLINA

PLATE 25.

TULIP TREE AT WOOLBEDING

PLATE 26.

TULIP TREE AT KILLERTON

In consequence of an accident Plate 27 cannot be ready for the present Volume, but will be issued with a later one.

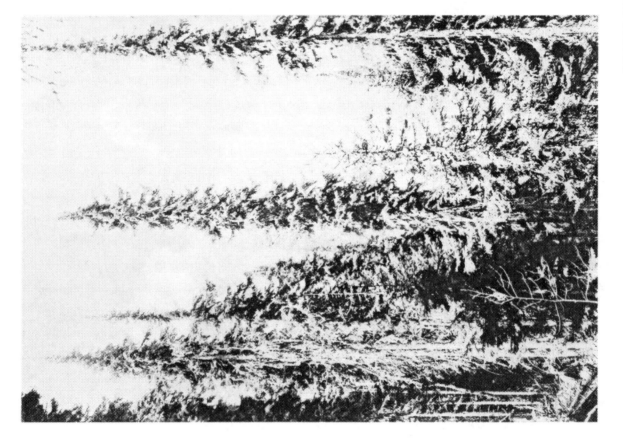

SERVIAN SPRUCE IN BOSNIA

PLATE 28.

PLATE 29.

BREWER'S SPRUCE IN AMERICA

PLATE 30.

MENZIES' SPRUCE AT BEAUPORT

JAPANESE YEW AT NIKKO

PLATE 31.

PLATE 32.

YEW AVENUE AT MIDHURST

YEW GROVE AT MIDHURST

PLATE 33.

PLATE 34.

IRISH YEW AT SEAFORDE

YEW AT TISBURY

PLATE 35.

PLATE 36.

YEW AT WHITTINGHAME

PLATE 37.

CRYPTOMERIA ELEGANS AT TREGOTHNAN

PLATE 38.

A

B

CRYPTOMERIAS IN JAPAN

Plate 39.

PLATE 40.

CRYPTOMERIAS AT IMAICHI

PLATE 41.

CRYPTOMERIAS AT NIKKO

PLATE 42.

CRYPTOMERIA AT HEMSTED

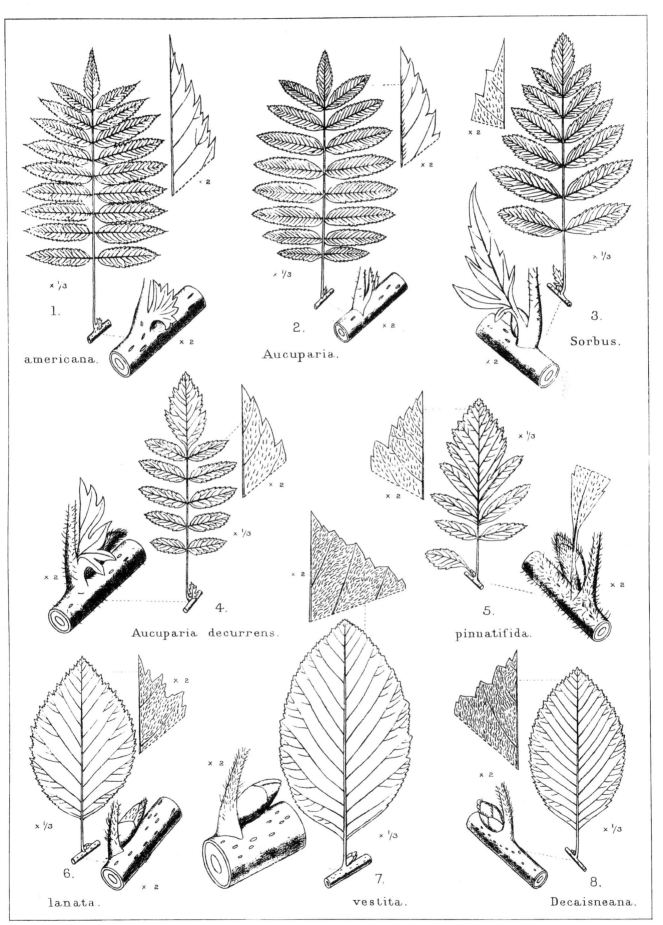

1.
americana.

2.
Aucuparia.

3.
Sorbus.

4.
Aucuparia decurrens.

5.
pinnatifida.

6.
lanata.

7.
vestita.

8.
Decaisneana.

PLATE 43.

PYRUS.

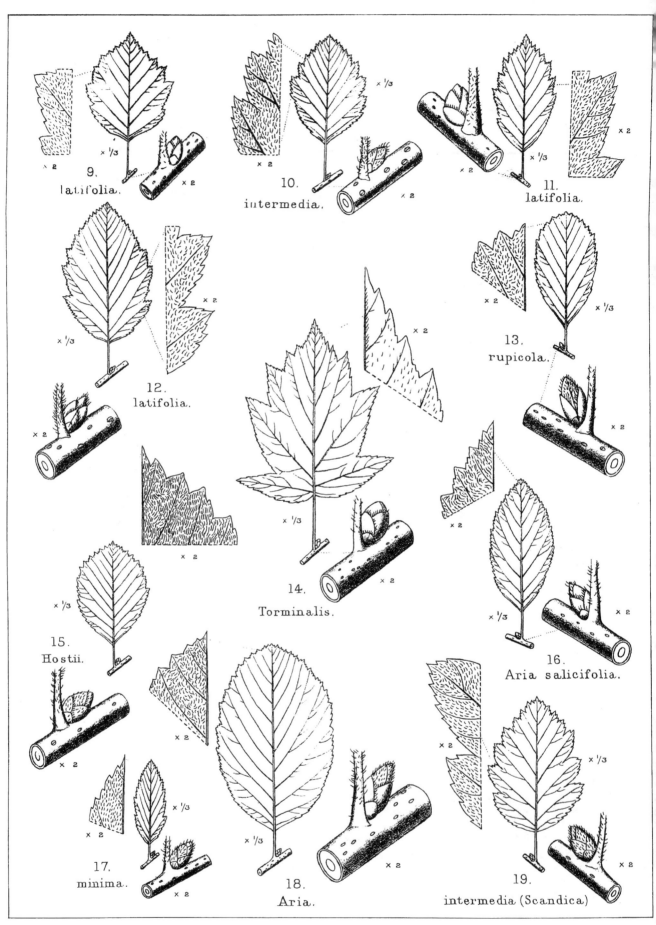

9.
latifolia.

10.
intermedia.

11.
latifolia.

12.
latifolia.

13.
rupicola.

14.
Torminalis.

15.
Hostii.

16.
Aria salicifolia.

17.
minima.

18.
Aria.

19.
intermedia (Scandica)

Huitt, del, Huth, lith.

PLATE 44.

PYRUS.

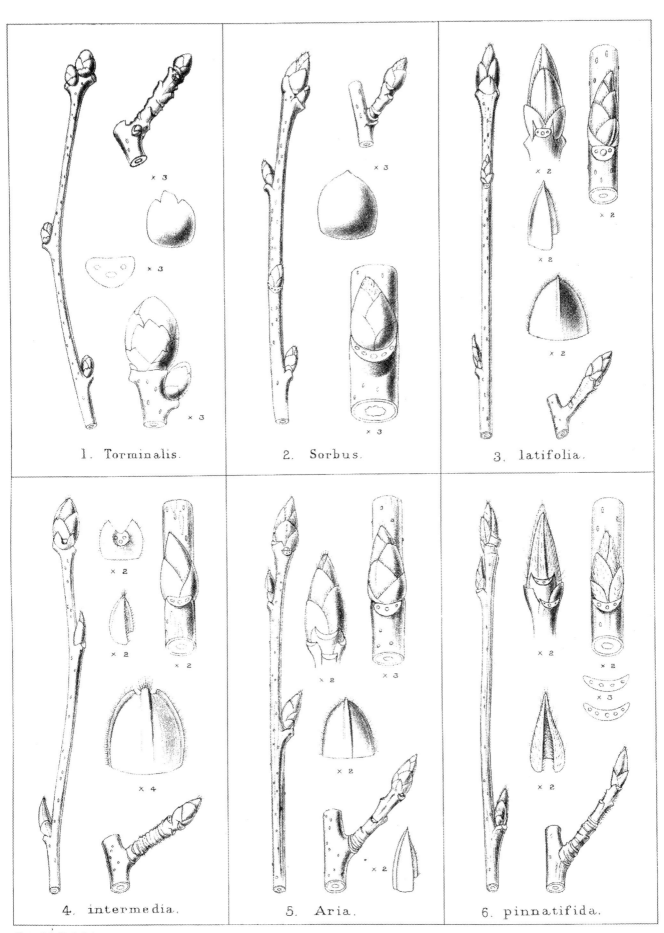

1. Torminalis.

2. Sorbus.

3. latifolia.

4. intermedia.

5. Aria.

6. pinnatifida.

Huitt, del. Huth, lith.

PLATE 45.

PYRUS

PLATE 46.

SORB TREE AT ARLEY

PLATE 47.

SERVICE TREE AT TORTWORTH

PLATE 48.

SERVICE TREE AT RICKMANSWORTH

PLATE 49.

SERVICE TREE AT OAKLEY PARK

PLATE 50.

SERVICE TREE AT SYON

PLATE 51.

WHITEBEAM AT CAMP WOOD

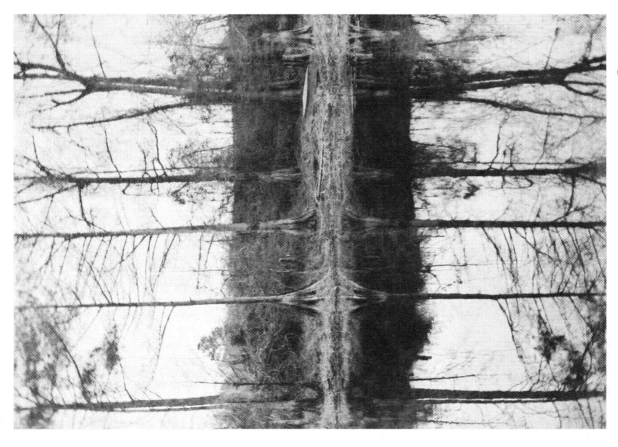

PLATE 52.

DECIDUOUS CYPRESS IN VIRGINIA

CYPRESS SWAMP IN NORTH CAROLINA

PLATE 54.

DECIDUOUS CYPRESS AT SYON

DECIDUOUS CYPRESS AT WHITTON

PLATE 56.

GIANT THUYA IN VANCOUVER'S ISLAND

GIANT THUYA AT FONTHILL

Plate 58 represents the trunk of what I believe to be the finest beech in Gloucestershire, which is known as "The Gladstone Beech," having been specially admired by the late Mr. Gladstone when staying at Cirencester House. It grows near Pinbury Park, formerly the seat of Sir Robert Atkyns, author of *The History of Gloucestershire*, and now the summer residence of Earl Bathurst. It measures about 110 feet high by 19 feet 9 inches in girth at the smallest part of the trunk, and though a very old tree, past its prime, is still a magnificent object. I am indebted for the negative to Mr. T. A. Gerald Strickland.